Use of Workers' Compensation Data for Occupational Safety and Health: Proceedings from June 2012 Workshop

David F. Utterback and Teresa M. Schnorr, Editors

Department of Health and Human Services
Centers for Disease Control and Prevention
National Institute for Occupational Safety and Health

May 2013

This document is in the public domain and may be freely copied or reprinted.

Disclaimer

Sponsorship of the Use of Workers' Compensation Data for Occupational Safety and Health Workshop by the National Institute for Occupational Safety and Health, the Occupational Safety and Health Administration and the Bureau of Labor Statistics does not constitute endorsement of the views expressed or recommendations for the use of any commercial product, commodity, or service mentioned. The opinions and conclusions expressed in the presentations and report are those of the authors and not necessarily those of NIOSH, OSHA or BLS. All conference presenters were given the opportunity to review and correct statements attributed to them in this report.

Recommendations are not final statements of NIOSH, OSHA or BLS policy or of any agency or individual involved. They are intended to be used in advancing the knowledge needed for improving worker safety and health.

Ordering Information

To receive documents or other information about occupational safety and health topics, contact NIOSH at:
Telephone: 1-800-CDC-INFO (1-800-232-4636)
TTY: 1-888-232-6348
Email: cdcinfo@cdc.gov
Or visit the NIOSH Web site at www.cdc.gov/niosh

For a monthly update on news at NIOSH, subscribe to NIOSH eNews by visiting www.cdc.gov/niosh/eNews.

DHHS (NIOSH) Publication No. 2013 – 147
May 2013

SAFER • HEALTHIER • PEOPLE™

Foreword

The Use of Workers' Compensation Data for Occupational Safety and Health Workshop was convened in June 2012 at the Frances Perkins Department of Labor Building in Washington DC. This was the second workshop that provided an opportunity for workers' compensation insurance industry organizations, public health practitioners and researchers, and government administrative agencies to discuss uses of workers' compensation data for public health issues.

The burden of occupational injuries, illnesses and fatalities is substantial. In the U.S. alone, costs are estimated at $250 billion annually (Leigh 2011). Tracking these costs and underlying hazards is essential for control of the economic and social burdens.

Workers' compensation insurance covers but a fraction of these costs, although nearly all employers are required by the individual state mandates to have policies. Seemingly, claims records would be available for each incident yet investigators report at this workshop and elsewhere that the records are incomplete.

Collaboration across the vested interests is needed to make workers' compensation data more suitable for research and surveillance purposes. In combination with other occupational safety and health resources, further utilization of workers' compensation data can help alleviate the burden of occupational injuries and illnesses in the U.S. and elsewhere.

John Howard, M.D.
Director
National Institute for Occupational Safety and Health
Centers for Disease Control and Prevention

Leigh JP (2011) Economic burden of occupational injury and illness in the United States. Milbank Q. 2011 Dec;89(4):728-72. doi: 10.1111/j.1468-0009.2011.00648.x.

Table of Contents

Introduction viii

Acknowledgements ix

Use of Workers' Compensation for Occupational Safety and Health: Opening Remarks 1

The Advantages of Combining Workers' Compensation Data with Other Employee Databases for Surveillance of Occupational Injuries and Illnesses in Hospital Workers 3

Safe Lifting in Long-Term Care Facilities, Workers' Compensation Savings and Resident Well-Being 7

Workers' Compensation versus Safety Data Use at the Veterans Health Administration: Uses and Weaknesses 17

Linking Workers' Compensation Data and Earnings Data to Estimate the Economic Consequences of Workplace Injuries 25

Workers' Compensation Costs in Wholesale and Retail Trade Sectors[1] 31

Linking Workers' Compensation and Group Health Insurance Data to Examine the Impact of Occupational Injury on Workers' and their Family Members' Health Care Use and Costs: Two Case Studies 41

Occupational Amputations in Illinois: Data Linkage to Target Interventions 47

The Role of Professional Employer Organizations in Workers Compensation: Evidence of Workplace Safety and Reporting[1] 51

Using Workers' Compensation Data to Conduct OHS Surveillance of Temporary Workers in Washington State 57

How WorkSafeBC Uses Workers' Compensation Data for Loss Prevention 63

Hitting the Mark: Improving Effectiveness of High Hazard Industry Interventions by Modifying Identification and Targeting Methodology 69

Injury Trends in the Ohio Workers' Compensation System[1] 73

Randomized Government Safety Inspections Reduce Worker Injuries with No Detectable Job Loss 79

Comparison of Data Sources for the Surveillance of Work Injury 81

OSHA Recordkeeping Practices and Workers Compensation Claims in Washington; Results from a Survey of Washington BLS Respondents 87

Completeness of Workers' Compensation Data in Identifying Work-Related Injuries 89

Another Method for Comparing Injury Data from Workers Compensation and Survey Sources 97

Using O*Net to Study the Relationship between Psychosocial Characteristics of the Job and Workers' Compensation Claims Outcomes 103

Impact of Differential Injury Reporting on the Estimation of the Total Number
of Work-Related Amputation Injuries 109

Exploring New Hampshire Workers' Compensation Data for its Utility in
Enhancing the State's Occupational Health Surveillance System 111

Using Workers' Compensation Data for Surveillance of Occupational Injuries
and Illnesses – Ohio, 2005–2009 117

Using an Administrative Workers' Compensation Claims Database for Occupational
Health Surveillance in California: Validation of a Case Classification Scheme
for Amputations 121

Describing Agricultural Occupational Injury in Ohio Using Bureau of Workers'
Compensation Claims 127

Use of Multiple Data Sources to Enumerate Work-Related Amputations in
Massachusetts: The Contribution of Workers' Compensation Records 133

Workers' Compensation-Related CSTE Occupational Health Indicators 135

The Effectiveness of the Safety and Health Achievement Recognition Program
(SHARP) in Reducing the Frequency and Cost of Workers' Compensation Claims 141

Comparison of Cost Valuation Methods for Workers Compensation Data 147

Development and Evaluation of an Auto-Coding Model for Coding Unstructured
Text Data Among Workers' Compensation Claims 153

Patterns in Employees' Compensation Appeals Board Decisions: Exploratory Text
Mining and Information Extraction[1] 157

Identifying Workers' Compensation as the Expected Payer in Emergency
Department Medical Records[1] 163

Utilizing Workers' Compensation Data to Evaluate Interventions and Develop
Business Cases 169

Gender, Age, and Risk of Injury in the Workplace 173

The Mystery of More Monday Soft-Tissue Injury Claims 179

Is Occupational Injury Risk Higher at New Firms? 183

Discussion of: Successes Using Workers' Compensation Data for Health Care
Injury Prevention: Surveillance, Design, Costs, and Accuracy. 185

Discussion of: The Total Burden of Work-Related Injuries and Illnesses: A Draft
White Paper Developed for the Workshop on the Use of Workers' Compensation Data 187

Discussion of: Workers' Compensation Loss Prevention: A White Paper for Discussion 191

Discussion of: Contingent Workers: Data Analysis Limitations and Strategies 193

Discussion of: Using Workers' Compensation Administrative Data to Analyze Injury
Rates: A Sample Study with the Wisconsin Workers' Compensation Division. 197

Discussion of: The Role of Leading Indicators in the Surveillance of Occupational
Health and Safety. 199

Final Workshop Discussion Group...201

State Health Agencies' Access to State Workers' Compensation Data: Results
of an Assessment Conducted by the Council of State and Territorial Epidemiologists, 2012..................203

Workshop Participants 209

Workshop Agenda..215

Poster Presentations 219

Introduction

David F. Utterback, PhD, Teresa M. Schnorr, PhD
National Institute for Occupational Safety and Health

Workers' compensation systems in the U.S. have grown complex since their initiation a century ago. All U.S. states (except Texas) require workers' compensation insurance coverage by nearly all employers. Each jurisdiction mandates that workers' compensation programs create reports for workplace injuries and illnesses and each state has an agency that collects at least a portion of these reports.

Standardized workers' compensation claims and program related information for a large portion of the states are also collected by industry organizations. Additionally, the workers' compensation insurance industry loss prevention programs generate records on employer risks and hazards. These resources on injuries, illnesses, hazards and other risks have yet to be fully utilized for occupational safety and health research and surveillance.[1]

The purpose for the June 2012 Use of Workers' Compensation Data for Occupational Safety and Health Workshop was to continue to explore ways in which workers' compensation information can be used for these purposes. The National Academies has called for greater use of surveillance data in order to identify priorities, focus resources and evaluate prevention program effectiveness.

Six white papers were drafted for the workshop and discussed in breakout groups. At the meeting, thirty-five poster and platform presentations described studies that utilized workers' compensation information while exploring limitations of these resources. These workshop proceedings contain summary articles for the presentations[2] plus notes from the discussion groups for the 6 white papers.[3]

The workshop was co-sponsored by the Bureau of Labor Statistics (BLS), Council of State and Territorial Epidemiologists (CSTE), International Association of Industrial Accident Boards and Commissions (IAIABC), National Council on Compensation Insurance (NCCI), National Institute for Occupational Safety and Health (NIOSH), Occupational Safety and Health Administration (OSHA), and the Washington State Department of Labor and Industries, Safety and Health Assessment for Research and Prevention (SHARP) program.

Continuing research and surveillance with workers' compensation resources can fill important gaps in our knowledge about workplace hazards and their impact on human health. Despite substantial differences among states, many public health and workers' compensation organizations are pursuing these opportunities (Appendix A). Everyone involved can help insure that the records for this complex industry are complete and accurate in order to maximize their potential use for protecting public interests.

---

[1]Proceedings from the first workshop are available at http://www.cdc.gov/niosh/docs/2010-152/
[2]Abstracts only appear for 5 articles that have been or are being published in peer-review journals.
[3]The white papers will be published in a peer-review journal.

Acknowledgements

We greatly appreciate the many contributors to the workshop planning and production. The workshop planning committee included Ben Amick, Les Boden, Rene Pana-Cryan, John Ruser, Teresa Schnorr, Glenn Shor, Harry Shuford, Barbara Silverstein, David Utterback, and Jennifer Wolf-Horejsh. The session moderators were Les Boden, Ted Courtney, Tish Davis, Bill Kojola, Steve Newell, and Tony Robbins. Breakout discussion groups were led by Christine Baker, Tim Bushnell, Linda Forst, John Mendeloff, Cameron Mustard, Rene Pana-Cryan, John Ruser, Teresa Schnorr, Harry Shuford, Marie Sweeney, Len Welsh, and Jennifer Wolf-Horejsh. The Occupational Safety and Health Administration hosted the meeting in the Department of Labor Frances Perkins Building in Washington, DC. We also thank those who presented at the workshop and all who engaged in the thoughtful discussions.

The workshop was co-sponsored by the Bureau of Labor Statistics (BLS), Council of State and Territorial Epidemiologists (CSTE), International Association of Industrial Accident Boards and Commissions (IAIABC), National Council on Compensation Insurance (NCCI), National Institute for Occupational Safety and Health (NIOSH), Occupational Safety and Health Administration (OSHA), and the Washington State Department of Labor and Industries, Safety and Health Assessment for Research and Prevention (SHARP) program.

Use of Workers' Compensation for Occupational Safety and Health: Opening Remarks

Bill Wiatrowski
Associate Commissioner at the Bureau of Labor Statistics

Good morning. I'm Bill Wiatrowski, Associate Commissioner at the Bureau of Labor Statistics. I am here to welcome you to the Department of Labor. It is nice to see several Department of Labor agencies represented here, all with an interest in worker safety and health.

As I was reviewing the agenda and participant list for this conference, I thought of the 1951 classic movie, When Worlds Collide. No, I don't think a stray planet is making a bee-line for this building. It's just that many of my worlds are coming together in this room. Consider:

- The Bureau of Labor Statistics is one of the sponsors of this workshop.
- There's an ongoing concern that BLS data undercount workplace injuries and illnesses; some of you are involved in that research.
- Some of the posters we'll see tomorrow involve automatic coding of injury narratives, a process BLS is attempting to learn more about.
- At least one of the presentations today uses data from the BLS Census of Fatal Occupational Injuries research file, a process my staff oversees.
- Other presentations look at workers' compensation costs and occupational characteristics, subjects included in another BLS program that I oversee, the National Compensation Survey.
- I note a number of friends in the audience from the National Academy of Social Insurance and several whom I work with on the NASI workers' compensation report.
- Other familiar names include those on BLS advisory committees, former employees, and long-time colleagues.

It's nice to see these worlds collide for a good purpose, to gain a better understanding of data on worker safety and health and to encourage good uses of those data to make safer workplaces.

Working for the Bureau of Labor Statistics, I would always say when it comes to data, more is better. You can never have too much high quality information. This idea has particular merit when compiling statistics on worker safety and health. We have experience using multiple data sources in the BLS Census of Fatal Occupational Injuries, or CFOI, which is in its 20th year. Details of fatal work injuries are gleaned from an average of 5 or 6 source documents, allowing us to confirm work relationship and identify many details about the worker, the employment, and the circumstances of the fatality.

In the past few years I've learned the importance of multiple data sources in other areas as well. While BLS data programs form the underpinning of our national injury and illness surveillance system, other data sources, including workers' compensation, can provide vital complementary information. To quote my former boss at the opening of the first of these conferences, "These data can supplement the BLS data with richer epidemiological information on the factors causing or associated with injuries and illnesses. They can provide better information about long run outcomes. And, these data may identify cases that are not captured by the BLS survey, perhaps because they are outside the Survey's scope."

The work we will hear about over the next two days will identify the type of information available through the workers' compensation system, but also explore challenges, such as variations across states and limits in scope. This workshop is about exploring the ways that workers' compensation data can add value to injury and illness prevention and ways that the limitations of these data can be overcome. My thanks to the organizers, presenters, and participants. I want to say a special thank you to my colleagues John Ruser and Eric Sygnatur for the many hours they have devoted to making this workshop a success.

I know that all of us will leave here with a deeper understanding of how workers' compensation data can achieve our joint mission to protect workers. I look forward to the conversation. Thank you.

The Advantages of Combining Workers' Compensation Data with Other Employee Databases for Surveillance of Occupational Injuries and Illnesses in Hospital Workers

Pompeii LA§, Dement JM,* Lipscomb HJ*, Schoenfisch A*, Myers D§, Østbye T.*

§University of Texas Health Sciences Center at Houston, *Duke University Medical Center

Introduction

The nature and setting of work carried out by hospital employees necessitates ongoing surveillance of work-related exposures and health outcomes.

Methods

The Duke Health and Safety Surveillance System (DHSSS) was developed in 2001 by Dement et al. (2004) as part of a NIOSH funded study aimed at improving the surveillance of work-related injuries and illnesses incurred by healthcare workers. The DHSSS is populated with occupational health data for healthcare workers employed in the Duke University Health System (DUHS) which includes a tertiary care medical center, two community hospitals, and their affiliated onsite and offsite clinics. To date, the DHSSS includes 14 years of occupational health data (1997 through 2010) on more than 20,000 healthcare workers. Workers' compensation data are included in the DHSSS and are linked at the individual employee level to other databases (Figure 1.0).

Workers' compensation data are designed for administrative purposes and their linkage to other data sources increases their utility with regard to examining occupational injury and illness risk, as well as evaluating the effectiveness of targeted prevention strategies. Human resources data are the core of the DHSSS. They include demographic and employment information on all DUHS workers and are updated annually. In addition, they include employees' number of hours worked per week and total months employed per year, which are used to construct full-time equivalent (FTE) measures as an estimate of time at risk. The FTE is essential for estimating workers' time at risk, and for calculating standardized rates of injury necessary for making comparisons across groups. The DHSSS includes numerous occupational health databases such as worksite health and wellness programs (e.g., Health Risk Appraisals), blood and body fluid exposures (e.g., NaSH data), and private health insurance claims (outpatient, inpatient, psychiatric, pharmacy), to name a few. The DHSSS includes a Job Exposure Matrix (JEM) where newly hired employees, and existing employees who change jobs within DUHS, are categorized for potential workplace exposures based on their job title and work department at the outset of their employment. The JEM includes more than 35 exposure categories such as blood and body fluid, hazardous drugs, noise exposure, and tuberculosis. For each category workers receive a code based on their potential level of exposure. For example, upon hiring or job change within the institution, all workers are coded for their potential exposures to tuberculosis: 1) no exposure, 2) direct patient care activities, 3) high-risk patient activities, or 4) works with non-human primates.

Annually, these databases are sent to an external company responsible for linking them at the individual worker level. Following HIPAA compliance, the data are then de-identified and uploaded into the DHSSS. Workers included in the system for more than one year have an individual line item for each year of their employment (or "report year" from 1997 through 2010) which consists of data pertaining to them that exists in each of the linked databases (Table 1).

Findings

This robust dataset has allowed us to examine a range of occupational health issues across DUHS occupational groups, departments and hospitals. Linked WC and HR data revealed that nurses' aides, housekeepers and dietary staff had the highest rates of musculoskeletal injuries over a 7-year time period, and that smaller workgroups, such as morgue technicians, patient transporters and skilled craft workers had higher than expected injury rates (Pompeii et al., 2007). These same data were used to examine the effectiveness of the implementation of patient handling equipment, and accompanying workplace policy, on musculoskeletal injuries and their associated costs across two hospitals over a 13-year time period (Schoenfisch et al., 2012; Lipscomb et al., 2012). Workers' compensation and HR data were used to calculate rates of patient handling injuries, as well as lost and restricted workdays among direct patient care providers. Lagged analyses were conducted to address the possible delayed impact of the intervention, given the time needed for hospital inpatient units to implement and train workers, as well as time needed for the adoption of this intervention. A significant protective effect was observed in the risk of patient handling-related injury among workers in one of the two study hospitals immediately following the intervention (RR: 0.56; 95% CI: 0.36, 0.87), with an increase in this protective effect as the lag time was examined at 6, 12 and 18 months post-intervention. The additional linkage of private health insurance data furthered these analyses by considering possible cost-shifting of musculoskeletal claims filed by workers during this study period.

In a separate analyses, WC and Health Risk Appraisal (HRA) data from the hospital's health and wellness program were used to examine associations between workers' body mass index (BMI) and work-related injury claims from 1997 through 2004 (Østbye et al., 2007). A linear association between increasing BMI categories and WC claim rates was observed. Compared to workers in the recommended BMI range (18.5-24.9 kg/height2), workers in the highest category ($\geq$ 40 kg/height2), had significantly more WC claims (5.5 vs. 11.7/100 FTEs), lost workdays (41.0 vs. 183.6/100 FTEs), and medical claims costs ($7,109 vs. $51,091/100 FTEs), respectively. The nature of injury most associated with higher BMI included sprains/strains and pain/inflammation, in addition to claims where the cause was coded as repetitive motion.

Our most recent analyses involved the assessment of the DHSSS at capturing workplace violent events where the worker was physically or verbally assaulted by a hospital patient or visitor (Type II Violence). Using WC and HR data, Rodriguez-Acosta et al. (2010) reported 1.7 physical assaults incurred by nursing staff from patients per 100 nursing FTEs. These analyses were expanded to examine these data from 2004 through 2009, and to assess the capturing of these types of events through other surveillance databases including the OSHA Log and the hospital's online voluntary Safety Reporting System. These analyses revealed that 484 Type II violent events were identified in this time period in at least one of these three data sets, with all of the events being patient-perpetrated. Rates were higher among male and Black workers, while older workers and those with greater work tenure had lower rates. Work groups identified as having higher rates included public safety workers (e.g., police, security guards), nurses' aides and nurses, and those working in psychiatry, police/transportation, float pool, neurology and the intensive care units. While WC data provided descriptive information about physical assaults that resulted in injury, Type II violent events that involved verbal and/or physical threats, assaults not resulting in an injury, and visitor-perpetrated events were not captured in the DHSSS. Furthermore, the voluntary Safety Reporting System was not as effective at capturing these types of events as expected, and details about circumstances surrounding events were sparse. Findings from these analyses, as well as those from previously published hospital-based Type II violence studies (Pompeii et al., in review), will be used to enhance the existing DHSSS to foster a more thorough capturing of these events.

Conclusion

While WC data provide work-related injury and illness information, they are limited in their utility, necessitating their linkage with other data sources. As summarized here, WC data are greatly enhanced when combined with HR data which allow us to define a cohort of workers, their demographic characteristics, and measures of hours worked which can be used to estimate their time at risk. The combination of WC and HR data is not uncommon in occupational epidemiology studies, but the linkage of WC with HRA, private health insurance, and online voluntary reporting systems data is, illustrating the broad utility of the DHSSS for examining work-related health issues. In addition to database linkage, the assessment of this cohort has been strengthened by the numerous years of data that includes information on more than 20,000 workers. Analyses of this large cohort revealed workgroups who were at risk for injury that have not been previously identified in hospital-based observational studies. A significant advantage of the DHSSS is that it allows us to examine rates of injury and illness over time within and across workgroups, departments and hospitals. The longitudinal nature of the data has been instrumental in examining the effectiveness of workplace interventions aimed at reducing the risk of work-related injury. This surveillance system is not without limitations, however. Our recent assessment of the System's ability to capture workplace violent events revealed areas that need improvement, including the need for more contextual details surrounding injury-related events. Without this, our ability to develop targeted prevention strategies is limited. This paucity of information is not isolated specifically to the reporting of workplace violent events, but is one we have faced when examining other occupational health issues, such as details surrounding patient handling-related injuries and the use of patient handling equipment. Through the use of focus groups and inpatient unit walk-through surveys, we learned about the barriers and promoters to adopting patient handling equipment (Schoenfisch et al., 2011) that were not provided in the WC data. The administrative nature of the WC system is not designed to capture circumstances of work-related events, and the DHSSS can be enhanced to fill this gap.

The DHSSS is a comprehensive data repository of numerous, linked databases pertaining to workers, their work environment and their health. This system will continue to be expanded and enhanced for purposes of increasing its utility for identifying workers at risk for injury and illness, and for developing and evaluating targeted prevention strategies.

References

Dement JM, Pompeii LA, Østbye T, Epling C, Lipscomb HJ, James T, Jacobs MJ, Jackson G, Thomann W. [2004]. An integrated comprehensive occupational surveillance system for health care workers. *Am J Ind Med* 45(6):528-38.

Lipscomb HJ, Schoenfisch AL, Myers DJ, Pompeii LA, Dement JM. [2012]. Evaluation of direct workers' compensation costs for musculoskeletal injuries surrounding interventions to reduce patient lifting. *Occupational and Environmental Medicine*. 69(5):367-72.

Østbye T, Dement JM, Krause KM. [2007]. Obesity and workers' compensation: results from the duke health and safety surveillance system. *Arch Intern Med* 167(8):766-773.

Pompeii LA, Lipscomb HJ, Dement JM. [2008]. Surveillance of musculoskeletal injuries and disorders among a diverse cohort of workers at a tertiary care medical center. Am J Ind Med. 51(5): 344-356.

Pompeii LA, Dement JM, Schoenfisch AL, Lavery AM, Souder M, Smith CD, Lipscomb HJ. Reported perpetrator, worker and workplace characteristics associated with Type II violence in the hospital setting: A review of the literature and existing occupational injury data. *Journal of Safety Research*. [In Review].

Rodriguez-Acosta RL, Myers DJ, Richardson DB, Lipscomb HJ, Chen JC, Dement JM. [2010]. Physical assault among nursing staff employed in acute care. Work. 35(2):191-200.

Schoenfisch AL, Myers DJ, Pompeii LA, Lipscomb HJ. [2011] Implementation and adoption of mechanical lift equipment in the hospital setting: The importance of organizational and cultural factors. Am J Ind Med. 54(12):946-54.

Schoenfisch AL, Lipscomb HJ, Pompeii LA, Myers DJ. Musculoskeletal injuries and disorders among hospital patient care staff before and after implementation of patient lift and transfer equipment. Scandinavian Journal of Work, Environment and Health, Early View: http://www.sjweh.fi/show_abstract.php?abstract_id=3288

Table 1. DHSSS Example of Annual Data Update

ID	REPORT YEAR	OCCUPATION	HIRE YEAR	TERM YEAR	YEARS EMPLOYED	AGE GROUP	RACE	GENDER	FTE
739	2004	nurse aide	2004		3. 10-15	25-34	BLACK	MALE	1
739	2005	nurse aide	2004		3. 10-15	25-34	BLACK	MALE	1
739	2006	nurse aide	2004		3. 10-15	25-34	BLACK	MALE	1
739	2007	nurse aide	2004		3. 10-15	25-34	BLACK	MALE	1
739	2008	nurse aide	2004		4. 16-20	35-44	BLACK	MALE	1
739	2009	nurse aide	2004		4. 16-20	35-44	BLACK	MALE	1
739	2010	nurse aide	2004		4. 16-20	35-44	BLACK	MALE	1
150	1997	pt transporter	1977	2002	5. >20	45-54	WHITE	FEMALE	1
150	1998	pt transporter	1977	2002	5. >20	45-54	WHITE	FEMALE	1
150	1999	pt transporter	1977	2002	5. >20	45-54	WHITE	FEMALE	1
150	2000	pt transporter	1977	2002	5. >20	45-54	WHITE	FEMALE	1
150	2001	pt transporter	1977	2002	5. >20	45-54	WHITE	FEMALE	1
150	2002	pt transporter	1977	2002	5. >20	>=55	WHITE	FEMALE	0.9

ID=Unique Worker ID; Report Year=Year of Employment; FTE=full-time equivalent, Term Year=Termination Year

Figure 1.0: Duke Heath and Safety Surveillance System (DHSSS)

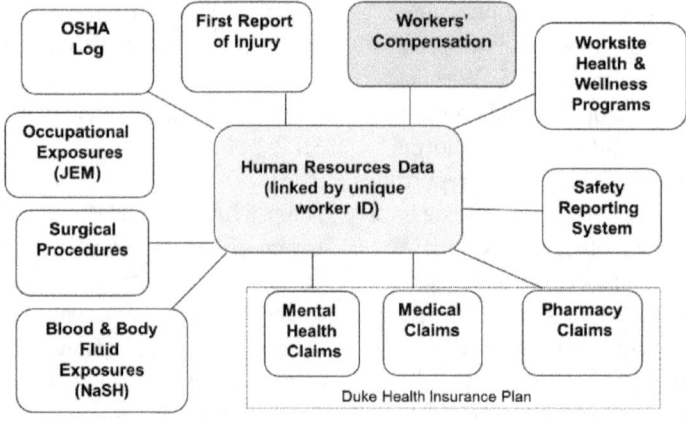

Safe Lifting in Long-Term Care Facilities, Workers' Compensation Savings and Resident Well-Being‡

Patricia W. Gucer, PhD§, Tanya Restrepo* MBA, Frank Schmid*, Harry Shuford*, PhD, Marc Oliver§, RN, MPH, Joanna Gaitens§, Phd., Dr. habil., Chun J. Shyong*, BA and Melissa A. McDiarmid§, MD, MPH, DABT

Introduction

We assessed the relationship between workers' compensation claims frequency and cost data, gathered originally for administrative purposes, and a) safe lifting programs and b) the availability of powered mechanical lifts (PMLs), measured during a nationwide survey of directors of nursing (DONs) between November, 2007 and May, 2008. We also assessed the relationship between safe lifting programs and lift availability with Centers for Medicare and Medicaid Services (CMS) resident well-being outcomes.

Safe Lifting and the Caregiver

Incidence of serious musculoskeletal (MSD) Injuries (with days away from work) is very high among nursing aides (232 per10,000) compared to private industry (33 per 10,000) and high even when compared to construction laborers (87 per 10,000) (BLS 2008). The causes of this very high injury rate are the forces on caregivers' musculoskeletal systems as they lift and transfer residents (Marras, Davis et al. 1999; Marras 2000). A body of research has shown that these injuries can be reduced by providing powered mechanical lifts (PMLs) in the context of a safe lift program (Brophy, Achimore et al. 2001; Evanoff, Wolf et al. 2003; Collins, Wolf et al. 2004). Here we uncouple the impact of a safe lifting program from the availability of powered mechanical lifts as we look at lift-related workers' compensation claims frequency and costs.

Safe Lifting and the Long-Term Care Resident

Since the resident is a participant in the lift process and vulnerable should there be a mishap, it makes sense that prevention of a caregiver injury during a lift would benefit the resident also. In a safe lift with proper use of a PML, the resident would be less likely to be dropped. Also, the use of PMLs may remove barriers to resident mobility, thus preventing such things as pressure ulcers, bedfastness, use of physical restraints and use of antipsychotic medication (without a diagnosis of psychosis). We also bear in mind that any resident mobility (as opposed to immobility) also carries risk for falls and fractures.

While the literature on resident benefits from lift use is sparse, Nelson and colleagues have examined the links between a comprehensive safe lift program and resident outcomes (Nelson, Collins et al. 2008) and found some benefits to resident alertness and affect.

The sit-stand lift (Figure 1) may offer particular benefits, since its use requires more effort (strength and balance) from the participating resident than does use of the full lift (Figure 2) in which the resident exercises none of his own muscle or balancing power to effect the lift.

---

§Occupational Health Program, University of Maryland School of Medicine
*National Council on Compensation Insurance
‡ Work funded by the Commonwealth Foundation and accepted for publication in the Journal of Occupational and Environmental Medicine

To look at the relationship between safe lifting and these two population outcomes, we joined safe lifting data from a survey of long-term care Directors of Nursing to two outside data sets:
- National Council on Compensation Insurance (NCCI) workers compensation claims and cost data (to assess caregiver outcomes).
- Centers for Medicare and Medicaid Services (CMS) facility level data on resident well-being (to assess resident outcomes).

Methods
DON Respondents
Two hundred seventy-one DONs in 23 states responded to a survey conducted by the Occupational Health Program at the University of Maryland containing questions about their powered mechanical lift (PML) inventories, resident census and characteristics of their safe lifting programs. A list of CMS certified LTC facilities was obtained (>7500) and NCCI matched the facilities on that list to a database containing claims information that yielded 656 facilities, just under a 10% match. All were invited to participate. DON surveys were conducted over the phone, on line or through the use of written mail in questionnaires. Surveys were conducted between November 2007 and May 2008

Survey variables from the DON survey
Variables obtained from the survey used to predict worker and resident outcomes included:
- Total PMLs per 100 residents 2005, 2006, 2007
- Sit-stand lifts per 100 residents 2005, 2006, 2007
- Full lifts per 100 residents 2005, 2006, 2007
- Resident census 2005, 2006, 2007
- Safe Lift Index (SLI) which was derived as the average z score of 11 items measuring facility policies, attitudes and practices regarding safe lifting including questions on training, lift need identification, use of powered mechanical lifts, accountability for failure to use lifts, and DON preferences for and perceived barriers to lift use. *Cronbach's alpha* was 0.749

Variables to determine the quality and duration of the safe lift program were also collected:
- Adequacy of safe lift policy ass assessed by a single question in the DON survey
- Time safe lift policy was in place

Workers' Compensation Variables provided by NCCI including the following:
Frequency outcome variable which reflects all claims due to lifting in nursing-related class codes (medical only and lost-time) per full –time equivalent worker at an annual rate for the years covered in the survey.

Cost outcome variable is the total medical and indemnity paid losses from the claims due to lifting in nursing-related class codes divided by exposure.

Statewide frequency
This variable is used to control for state differences.

Resident well-being variables collected from CMS included:
Outcome variables - facility level resident quality indicators (QI/QM) from the CMS "percent triggered" file, indicating the percent of facility residents who are determined by the facility to fall above a predetermined "triggering" threshold score on each indicator below:
- Use of physical restraints
- Use of Antipsychotic drugs without a diagnosis of psychosis
- Bedfastness
- Residents at high risk for and have pressure ulcers
- Falls
- Fractures

Ownership structure variable to control for differences in for-profit, not-for-profit, and government owned facilities, obtained from the CMS Nursing Home Compare website data.

Analyses
Workers' Compensation Outcomes
Models (claims frequency and costs) are repeated measurement multilevel Tobit models with random effects at the unit of measurement (LTC facility) and fixed effects on the level of ownership type. (Tables 1 and 2)

Resident Well-Being Outcomes
We examined resident well-being mean values stratified by safe lift index for 2005, 2006, and 2007. See Table 3.

We examined nonparametric Spearman correlations between resident well-being outcomes and lift type, for 2005, 2006 and 2007, presented in Table 4

We used cross sectional generalized estimating equations (xtgee) (STATA 11) to examine the relationships between resident well-being outcomes and a) sit-stand PMLs and b) full-lifts per 100 residents, c) the SLI, adjusting for year and census. This information is presented in Table 5.

Results
Workers Compensation Outcomes
Claims Frequency
See Table 1. Higher values of the safe lift index are significantly associated with lower values for workers' compensation claim frequency (p<= .01). Also, while not reaching statistical significance, we note the beneficial impact on workers compensation claims frequency of the number of available lifts per resident.

We observed that for a one standard deviation increase in the SLI, a 49% reduction in claims frequency is obtained.

Costs
See Table 2. Higher values of the safe lift index are associated with lower workers compensation costs (p=<.05).

Further analysis showed that a one standard deviation increase in the safe lift index is associated with a 33% reduction in total facility compensation costs.

Resident Well-Being Outcomes
Outcomes stratified by SLI
Please see Table 3. In all six outcomes, the resident well-being means are lower (better) in the high SLI category than the low. However, we see no significant differences in these mean outcomes by SLI category (low/high).

Correlations between PML availability and resident outcomes over three years. Please see Table 4. Seventeen of the 18 correlations between total lifts per 100 residents and outcomes are statistically significant, with correlations of all resident outcomes but falls and fractures in the expected direction, that is, better resident outcomes are observed when more lifts are available. We see similar patterns when we look separately at full and sit-stand lifts. Associations are generally negative (yielding better resident outcomes) between lift availability and restraint use, bedfastness, antipsychotic drug use and pressure ulcers. These beneficial associations appear stronger for sit-stand lifts with 18 vs. 6 comparisons reaching statistical significance. The associations between lift availability and falls and fractures are positive, reflecting the residents' exposure to the risk of falling when they are no longer immobile.

Multivariate analyses of resident outcomes by safe lift predictors. Please see Table 5. Overall, we see that safe lift predictors have a generally beneficial effect on resident well-being outcomes, adjusting for the size of the facility and year. This is true for full and sit-stand lifts which are both associated with a critically important resident well-being outcome - fewer pressure ulcers among high risk residents. In addition, sit- stand lifts are associated with less bedfastness. We also see that sit- stand lifts are slightly associated with falls and fractures. The safe lift index however, which integrates comprehensive elements of a safe lift program, is associated with a decline in resident falls.

Discussion

Although the impetus for safe lift practices and lift assist device usage in LTC was originally meant to stem the high rates and costs of caregiver injury, we have shown in this study that they are also associated with benefits to residents in LTC. For four CMS-derived mobility-related resident well-being outcomes, we observed improvements for residents (two statistically significant after accounting for adjustment variables) as a function of the lift number. The positive associations with falls may reflect to some degree an inherent fall risk associated with resident mobility. However, resident falls decline with better safe lift programs which include comprehensive safe lift strategies, policies and worker training, as measured by the Safe Lift Index.

Reducing workers compensation claims and costs also depended not just on lift numbers, but on these integrated facility safe lifting characteristics.

Thus a comprehensive safe lift culture in LTC was seen to benefit both workers and residents in an integrated system of safety.

References

BLS (2008). "Incidence rate and number of musculoskeletal disorders, selected occupations, 2008." Fatal Occupational Injuries and Nonfatal Occupational Injuries and Illnesses, 2008Bureau of lLabor Statistics, 2008(<http://www.bls.gov/iif/oshwc/osh/os/oshs2008.pdf>).): 48.

Brophy, M. O. R., L. Achimore, et al. (2001). "Reducing incidence of low-back injuries reduces cost." *Am. Ind.Hyg. Assn. J.* 62(4): 508-511.

Collins, J. W., L. Wolf, et al. (2004). "An evaluation of a "best practices" musculoskeletal injury prevention program in nursing homes." *Injury Prevention* 10(4): 206-211.

Evanoff, B., L. Wolf, et al. (2003). "Reduction in injury rates in nursing personnel through introduction of mechanical lifts in the workplace." American Journal of Industrial Medicine 44(5): 451-457.

Marras, W. S. (2000). "Occupational low back disorder causation and control." *Ergonomics* 43(7): 880-902.

Marras, W. S., K. G. Davis, et al. (1999). "A comprehensive analysis of low-back disorder risk and spinal loading during the transferring and repositioning of patients using different techniques." *Ergonomics* 42(7): 904-926.

Nelson, A., J. Collins, et al. (2008). "Link between safe patient handling and patient outcomes in long-term care." *Rehabilitation Nursing* 33(1): 33-43.

Table 1. Workers' Compensation Claims Frequency Tobit Model

Explanatory Variable	Coefficient	Standard Error	Significance
(intercept)	-0.1634	1.1173	
Lifts per Resident	-0.0741	0.0536	
Safe Lift Index	-0.1733	0.0399	***
State Frequency	2.731	0.8335	***
For Profit	-0.8190	0.435	*
Government	-0.2378	0.6941	
(Log of scale parameter)	0.7410	0.0598	***

No. of observations: 317
No. of positive observations: 216
Log Likelihoods: -542.6 (model); -655.1 (intercept only)
  Chi-squared (58.8): 225.13***
Analysis of variance of safe lift program:
  Chi-squared (2) 26.45***
Analysis of variance of ownership structure:
  Chi-squared (2) 4.19
Note ***,**,* significance at 1%, 5% and 10% levels, respectively

Table 2. Workers' Compensation Total Costs Tobit Model

Explanatory Variable	Coefficient	Standard Error	Significance
(intercept)	-0.1465	0.2283	
Lifts per Resident	-0.0101	0.0110	
Safe Lift Index	-0.0209	0.0082	**
State Frequency	0.4995	0.1703	***
For Profit	-0.1744	0.0887	**
Government	-0.1713	0.1430	
(Log of scale parameter)	-0.8347	0.0549	***

No. of observations: 317
No. of positive observations: 213
Log Likelihoods: -196.3 (model); -275.9 (intercept only)
  Chi-squared (56.6): 159.21***
Analysis of variance of safe lift program:
  Chi-squared (2) 13.91***
Analysis of variance of ownership structure:
  Chi-squared (2) 6.15**
Note ***,**,* significance at 1%, 5% and 10% levels, respectively

Table 3. Resident well-being outcomes by category of safe lift index: 2007

Safe Lift Index, categorized into low and high*		Percent of facility residents who					
		Were physically restrained	Were given antipsychotic medication without a diagnosis of psychosis	Spent most of their time in bed or in a chair	Had pressure ulcers while at high risk	Fell in the past 30 days	Broke a bone in the past quarter
Low (n=126)	Mean	4.03	21.33	3.06	13.72	15.42	1.59
	SD	5.64	9.12	3.68	8.72	5.27	1.09
High (n=132)	Mean	3.72	20.51	2.49	12.45	14.49	1.51
	SD	5.31	10.54	3.27	6.63	4.32	1.09
Total (n=258)	Mean	3.87	20.91	2.77	13.07	14.94	1.55
	SD	5.47	9.86	3.48	8.43	4.84	1.09
Anova		p=.615	p=.507	p=.186	p=0.190	p=0.121	p=0.538

* cut is by mid point of the index frequency values in the safe lift index (SLI)

Table 4. Correlations between PML availability and resident outcomes over three years

Percent of Facility Long-term care residents who:	Total Lifts[a]			Full Lifts			Sit Stand Lifts		
	2005 N=233[b]	2006 n=247	2007 n=263	2005 N=234 total	2006 N=248	2007 N=264	2005 N=243	2006 N=251	2007 N=266
Were physically restrained	-.144*	-0.201**	-.234***	-.100	-.107	-.115	-.159*	-0.234***	-.265***
Received antipsychotic use w/o diagnosis of psychosis	-.240***	-.166**	-.183**	-.161*	-.072	-.066	-.240***	-.187**	-.213***
Spent most of their time in bed or in a chair	-.239***	-.224**	-.219***	-.138*	-.148*	-.077	-.238***	-0.193**	-.238***
Had pressure ulcers while at high risk	-.304***	-.261***	-.260***	-.214**	-.111	-.014	-.322***	-.280***	-.278***
Fell in the past 30 days	.232***	.169**	.126*	.176**	.040	.036	.217**	0.269***	.159*
Broke a bone in the last quarter	.209**	.157*	.086	.153*	.034	.040	.157*	.243***	.181**

Correlations are nonparametric, using Spearman tests of significance.

*prob <.05, **prob <.01, *** p >.001

[a] overhead lifts are not included in this analysis, only 3 facilities had overheads in 2007

[b] numbers decline in earlier years as some DONs declined to estimate numbers of lifts in previous years

Table 5. Significant associations[a] between resident well-being outcomes Safe Lift Predictors

Resident Well-Being Outcomes

	Physical restraint		Antipsychotic drug use		Bedfastness		Pressure ulcers		Falls		Broken bones	
	sign	p value	sign	p value	sign	p value	sign	p value	sign	p value	sign	p value
Predictor variables												
Sit stand PML/100 residents					neg	**	neg	***	pos	**	pos	***
Full lifts/100 residents							neg	*				
Safe lift index					neg	*			neg	*		
Adjustment variables												
Size (number of occupied beds)	pos	**										
Year	neg	*	neg	**	neg	*	pos.	*	neg	***	neg	***

* p<.05, ** p<.001, *** p < .000

[a] Results from generalized estimating equations (XTGEE in Stata 11), negative binomial distribution, robust SE, and auto regressive cirrelation among subjects (facilities)

Figure 1. Sit-Stand Lift

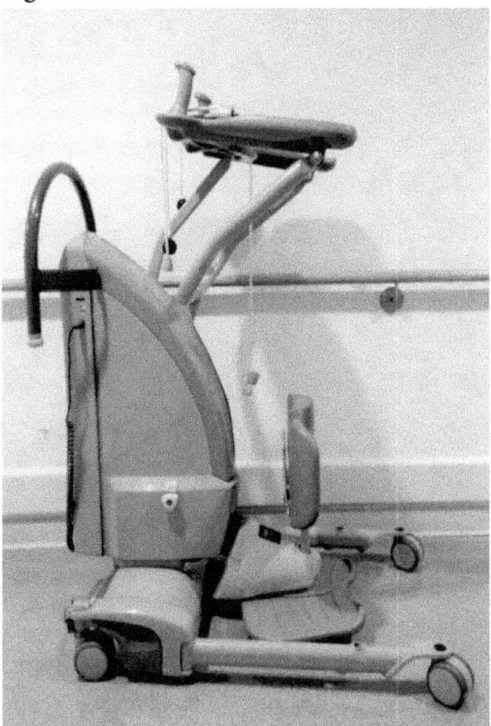

Figure 2. Full lift

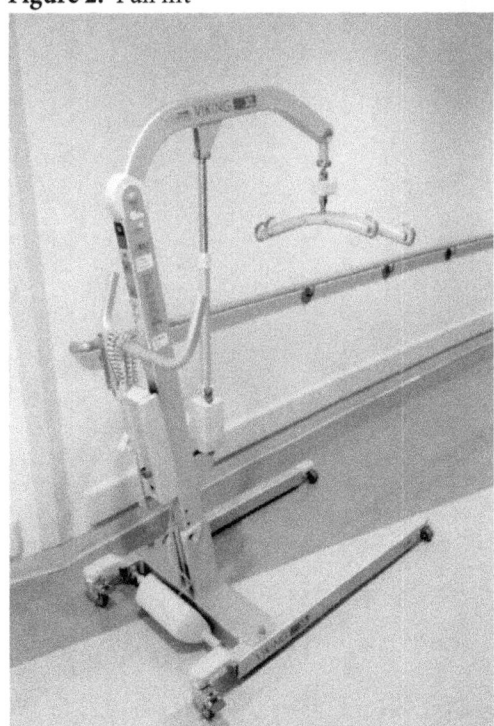

Workers' Compensation versus Safety Data Use at the Veterans Health Administration: Uses and Weaknesses

Michael Hodgson, M.D., M.P.H.
Veterans Health Administration

Background

In the mid 1990s, the Veterans Health Administration (VHA), one of three administrations in the Department of Veterans Affairs (VA), undertook a major restructuring of its systems, focused on quantitative performance management. Given its long –standing focus on appropriate use of data, it created a vision of aligned integrated data systems, with elements of safety management, workers compensation (WC), clinical employee occupational health, and hazard management integrated into a single unified system. A basic WC system did exist, as a franchise fund, supporting both VHA and several other Federal agencies. The national safety system, the Automated Safety Incident Surveillance and Tracking System (ASISTS), a safety management system, developed at one of the facilities, was expanded to support collaboration among all parties involved in injury management. These included the injured employee, the supervisor, and staff from WC, safety, and occupational health. This presentation will summarize the use of data for program management and evaluation in a large health care system using signature injuries as examples.

Patient handling, sharps, and assault represent signature injuries in the health care industry, the industry with the highest rate of nonfatal injuries in the U. S. workplace. Each of these presents complications or record-keeping. Patient manual handling injuries have no specific code within the old Bureau of Labor Statistics coding system; only a complex algorithm identifies them, and that approach has never been validated. Nevertheless, they represent approximately 15% of injuries within VHA's safety management system and over 40% of injuries to nursing personnel. The definition of assaults varies dramatically between law enforcement, safety, and healthcare usage and even between police, safety, and human resources within the same system. Within VHA those represent approximately 7% of all injuries. Together, assaults and manual handling injuries represent almost 80% of injuries to nursing personnel. Sharps injuries, with reporting requirements defined by 29 CFR 1910.1025, represent approximately 14% of safety incidence but less than 0.1% of worker's compensation events. The WC data system for VHA contains fewer than 35% of the overall injuries reported to the safety system, and in approximately 35% of the events in the workers compensation system, the code for type of injury is missing. In 2004 ASISTS injury data were made available to all levels of the organization under the assumption that knowledge of local injury rates, and arising competition, would drive awareness and prevention behaviors. For Privacy Act reasons, access to the WC system is controlled more tightly.

VHA has used data as a core tool in its performance management not just in clinical management (Jha et al. 2003) but also for infrastructure operations. An example (Figure 1) presents the results of three workers compensation performance indices over the last years, as they came into use. VHA chose to rely on the White House performance metric, timely submission (defined as occurring within two weeks after injury) of WC claims (CA1/2 submission timeliness). The goal was revised upwards each year, as performance across Federal agencies improved, Beginning in 2009, additionally, VA tracked the number of individuals not at work who had work capacity. Beginning in 2011, timely submission of fiscal paperwork ("Department of Labor CA7 form submission timeliness", i.e., within one week) became a White House performance metric. Figure 1 presents a timeline of performance improvement on central WC metrics followed both by the Department of Labor and by VA and VHA management. The graph shows clear evidence of progress with measurement and feedback being associated with dramatically improved performance.

Importantly, the major improvements in performance metrics did not accompany parallel financial management improvement, however. Between 2000 and 2011, despite the above dramatic improvement, both the compensation chargeback[1] costs (i.e., wage and salary replacement for longer-term disability [CCBC] and chargeback medical costs [CBMC]), costs for testing and treatment, increased dramatically, by over 50%. A major reason for that discrepancy rests in the way the Federal system tracks short term disability costs.

Those costs, incurred during the first 45 days after injury, are formally called "continuation of pay" (COP) and are managed by individual agencies with a simple payroll coding change. The CCBC represent costs beginning after 45 days after injury. The actual figure changes frequently, until DOL OWCP makes a final determination whether an injury is really work-related. The Federal program is regulated under the Federal Employee Compensation Act, which contains far more stringent privacy protections then commercial workers compensation systems.

Although VHA has been a leader in the development and use of the electronic patient record, its human resources infrastructure has not benefitted from the same sustained strategic focus and support. The central reporting of COP Days requires multiple hand-entries, telephone and paper notes between three participants (supervisor, payroll clerk, WC specialist), and manual data transmission. Figure 2 illustrates the arising discrepancies for

Figure 1. VHA WC Program Management: 2000-2012.

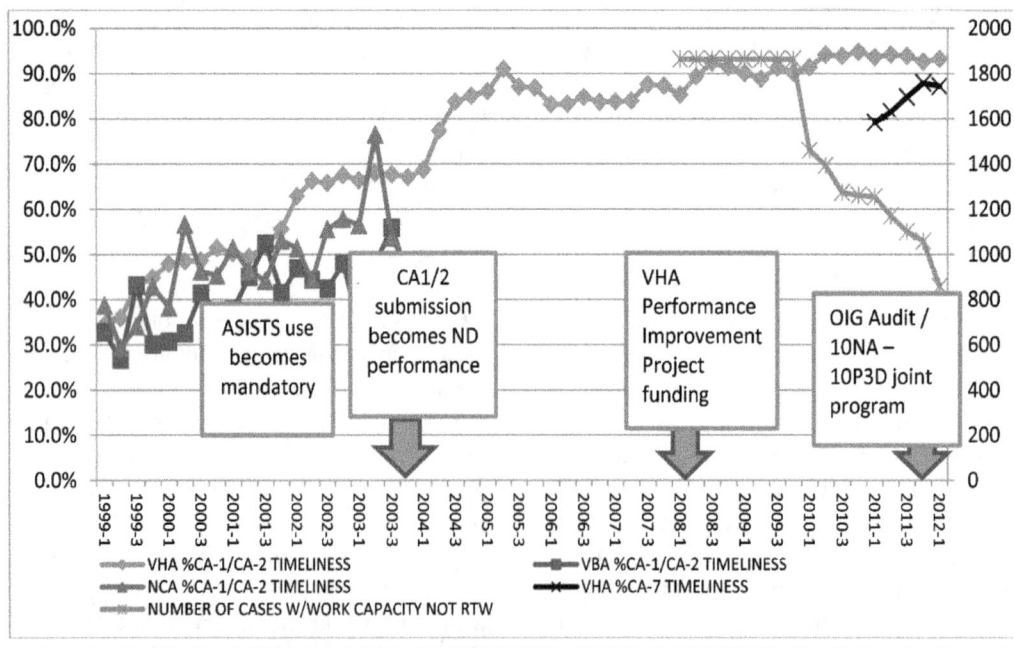

[1] The term "chargeback" refers to the process by which Occupational Workers Compensation Program (OWCP) bills employing agencies for their compensation costs, which are calculated on the basis of payments made from the Compensation Fund. By August 15th of each year, OWCP informs each agency of the amount expended on behalf of its employees from the Compensation Fund during the preceding fiscal year (which runs from July to June for chargeback purposes). The agency then either reimburses the Fund in the amount requested or budgets that amount for compensation purposes for the upcoming fiscal year. The process is described in FECA PM 5-0900).

the first quarter of the fiscal year 2012 for COP data, as reported through two different process streams. One stream provides information on the number of claims authorized by staff and is reported directly to the VHA WC program office. The second represents the figure that actually finds its way to the Governmental central accounting system and is reported to the various agencies. The national system provided a 38% under estimate of cost. This discrepancy occurs each quarter. The complexities of managing administrative data, with its deadlines and operational timing requirements, obviously influence data quality in operations and distinguish them from data quality in research settings. Practitioners and researchers must be aware of the context and environment in which data are collected.

Patient Handling Injuries

Patient handling represents one of the signature injuries in health care and has long represented an highly visible hazard. In recognition, the Tampa VAMC initiated a program in 1998. All patient manual handling injuries to nurses were examined, classified to a mechanism, and coded by preventability characteristics (Nelson et al. 2003). An expert panel was assembled to redesign the tasks.

Redesign of transfer tasks was pilot-tested and validated in a biomechanics lab in 2000 and underwent an initial field demonstration project (Nelson et al. 2006). Between 2001and 2003 the program was rolled out in highrisk (spinal cord injury, nursing home care) units under carefully controlled evaluation with detailed business case considerations (Siddharthan et al., 2005). Data from that roll-out project, collected under Institutional Review Board approval, with assurances of confidentiality, supported a strong business case. Beginning in 2005, VHA's Office of Public Health supported national implementation at individual, early adopting facilities and regional business units. National program cost modeling in 2006-2007 led to a formal budget proposal, funded for 2008 through 2011 at $200 million. The program consisted of technology (ceiling lifts, other equipment), unit peer safety leaders, and facility-level program managers together with additional program elements. Figure 3 presents the annual rate of manual handling injuries, as described in ASISTS, together with critical milestones. Injury rates continued to rise through 2006, because of increased patient acuity, gradually increasing patient obesity, and the aging workforce.

Figure 2. Continuation of Pay Hours: Concordance of VHA Internal Authorized Reporting with AITC/DFAS Reporting

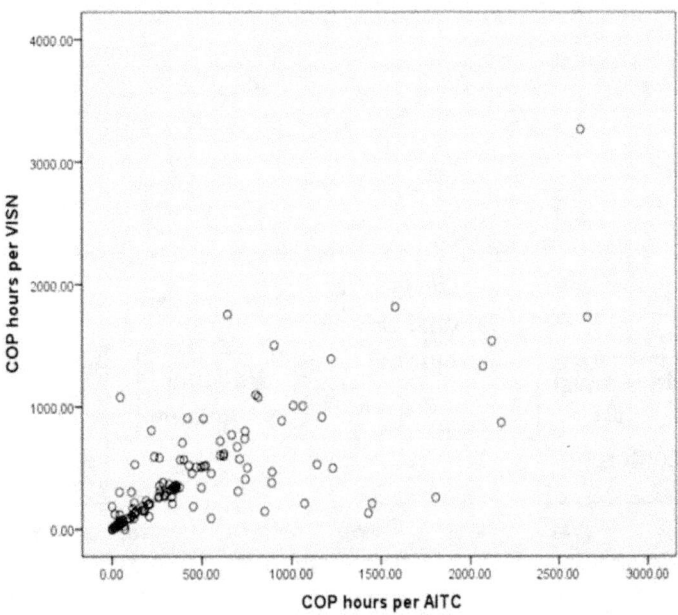

These rate changes, though encouraging, are misleading and incomplete in at least three ways. First, even this system [ASISTS] suffers from approximately 50% under reporting [Siddharthan et al., 2006]. Second, short-term disability costs (COP days) are not linked to any of the data systems, and neither ASISTS or WC-OSH-MIS provides short-term disability costs related to the injuries. For that reason, cost savings are unavailable at the national level. Third, the WC system, which uses BLS codes, has no specific manual handling category so that even long term disability costs are not precise and cannot be linked to injury cause without individual evaluation and pulling of each administrative record. Operational WC data, with old BLS codes, as used in this large system, therefore pose real problems of miscoding, misclassification, and under-estimation of costs even though some performance rate data suggest major improvements in injury frequencies. Justification of the national program on a cost basis would not have been possible using a business case development without a very formal scientific evaluation under Human Subjects Protection rules. For musculoskeletal injury tracking, therefore, the safety system has had some advantages.

Violence and Assaults

Similar issues arise with assaults, a second signature a hazard in the healthcare industry, as almost 60% of nonfatal assaults in the workplace occur in that industry. VHA initially developed systematic front-line worker protection programs in the late 1970s, Prevention and Management of Disruptive Behaviors [Lehman 1979], with awareness, personal safety skills, de-escalation techniques, and therapeutic containment strategies taught in four modules. In 2000, VHA undertook a national review and initiated major program shifts to reduce the threat of violence and to improve workplace safety, distinguishing patient-driven from coworker violence. After leadership training and a national survey [Hodgson et al. 2004], a national stand-down for violence prevention awareness occurred in 2002. The trainer network was rejuvenated over the next years, placing at least two at each facility, and training front-line staff [Mohr et al. 2011]. In 2004 in 2005, a Disruptive Behavior Committee was established at each facility to manage the "Patient Record Flag", an item visible across the country in the VHA's electronic medical record [Hodgson 2012]. A one-week on-site mini residency teaches the necessary threat assessment and

Figure 3. Time course of the national patient manual handling injury rate in relation to major program initiatives based on ASISTS

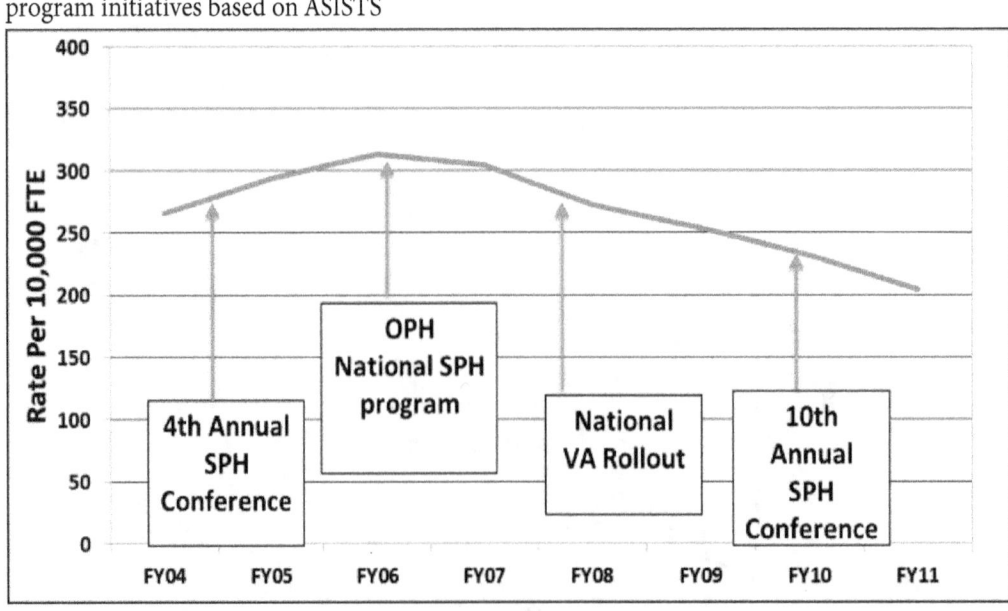

management skills. In response to the perceptions of coworker threats, VHA implemented the Civility, Respect, and Engagement of the Workforce [CREW] project [Belton 2007].

Still, measurement of success in reducing violence requires data. Data require both a system and definitions. The definitions of assault vary substantially between different systems. For example, in law enforcement verbal aggression falls under the definition of assaults; the distinction between assault and battery rests on touching. Most safety professionals consider assaults to require physical injury. In the world of human resources, hostile work environments, lateral violence, and other "looser" definitions are included. Figure 4 presents the annual rates of assaults in the context of VHA's major program initiatives since 2000. The specific question, then, is whether rates in safety systems and rates in workers compensation systems provide the same answer. We assembled rates at the facility level for three fiscal years, 2008, 2009, and 2010, from our 140 facilities. We explored correlation coefficients between those years for rates at the facility level comparing the safety and the workers compensation system. The results suggest a reasonable relationship, but that relationship changes dramatically from year to year. In 2008, the correlation coefficient was 0.25, clearly statistically significant ($p < 0.0051$), although only 6% of the variance was shared across facilities [R-square of 0.062] with 15, or 10.7% facilities missing data. In 2009, the correlation coefficient was 0.45, substantially higher with a proportionately larger, i.e., 20%, shared variance ($p<0.001$) and 13, or 9.3%, missing data. In 2010, the correlation coefficient was 0.648 ($p<0.001$), with almost half of the variability across facilities shared [R-square of 0.42] with 12 (8.6%) facilities missing data. These figures suggest that at least for something as dramatic as assaults relatively similar figures arise at the facility level. On the other hand, more detailed scrutiny provides very different answers. For example, in a 2001 survey, patients were perpetrators of over 85% of assaults on clinicians and on over 65% of assaults on non-clinical staff [Hodgson 2004] with the remainder resulting primarily from staff on staff violence. Recent

Figure 4. Time course of the national assault injury rate in relation to major program initiatives based on ASISTS

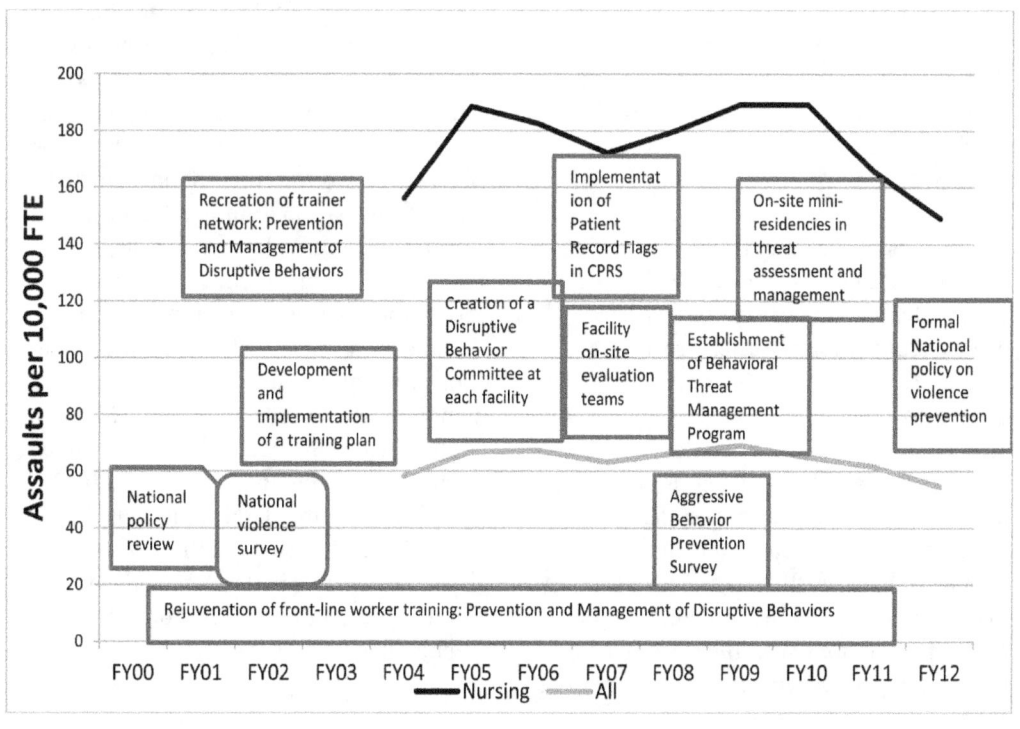

data [Dement 2004, Arnetz 2012] suggest a far higher proportion of staff driven assaults. In addition, detailed scrutiny of the approximately 6000 events in VHA's safety system, ASISTS, suggests that over 95% represent patient on provider attacks. Again, as with safe patient handling, somewhat similar results arise from workers compensation and a safety data, but the differences are worth exploring. Policy developers and scientists should know their organization, data systems, and criteria.

Blood Borne Pathogens Injuries
Finally, the third signature injury in healthcare, sharps, has generated its own recordkeeping requirement in the Sharps Injury Log, as required by 29 CFR 1910.1025. For a variety of reasons, over 95% of sharps injuries do not generate a formal, compensable, recordable workers compensation injury. Very few individuals lose more than a few hours of work time. And yet within the VHA system, sharps injuries represent approximately 14% of all injuries, upwards of 5000 injuries per year.

Summary
Clearly, workers compensation data have their uses. Still, users must be aware of the limitations. This presentation provided evidence for several such limitations; others, such as cost shifting, hidden costs, and enrollment criteria, are well known. Here four limitations are worth reiterating.

First, results and interpretations of workers compensation data differ from those arising from the use of safety data even within the same system. A primary concern is data quality, especially for researchers who are used to the accuracy and integrity of research-quality data.

Second, there are clear differences in rates derived from the two separate systems; the reason for that discrepancy remains unclear. Similarly, characteristics of the events captured in each system may differ, as described above for the perpetrator issue. The reasons for those discrepancies remain unclear, as reasons for reporting into two separate systems have not been studied.

Third, the issues of definitions must be emphasized. Some systems have very clear, operationally well-defined criteria for use. In others, the definitions are no less precise but differ substantially from the way they are used elsewhere. In yet others no formal definitions exist, and miscounting is likely.

Fourth, data can only be understood in their system's context. Users must understand the organization, reporting behaviors, definitions, incentives, and technical details of reporting systems.

In any case, the greater use, publication, and targeted dissemination would drive attention to system performance

Acknowledgements
The author wishes to acknowledge the extensive work of Charles Welch, PhD, in managing the ASISTS data and Sara Moreau for obtaining the assault rates in WC-OSH-MIS. The work on patient handling injury prevention would not have occurred without the tireless energy Audrey Nelson, Ph.D., R.N.; Mary Matz, M.S.P.H.; Gail Powell-Cope, Ph.D., R.N.; and the team of researchers at the Tampa Patient Safety Center of Inquiry and the national community of facility champions. The work on violence prevention benefited greatly from David Drummond, Ph.D.; Richard Reid, MSW; Lynn Van Male, Ph.; and Laurent Lehmann, M.D. The expansion of both the violence prevention / behavioral threat management and the safe patient handling programs occurred through the strong support of Lawrence Deyton, M.D., M.P.H. Eileen Coyne, R.N., M.P.P., led VHA's workers' compensation program from 2002 to 2012. Arnie Bierenbaum, M.S.E., Fran Murphy, M.D., M.P.H., and Susan Mather, M.D., M.P.H. played a major role in developing the data vision, initiating the various programs, and shepherding them over the years.

References

Arnetz JE, Aranyos D, Ager J, Upfal MJ. Development and application of a population-based system for workplace violence surveillance in hospitals. *Am J Ind Med.* 2011;54:925-34. doi: 10.1002/ajim.20984. Epub 2011 Jul 7

Belton LW, Dyrenforth SR. Civility in the workplace. Measuring the positive outcomes of a respectful work environment. *Healthc Exec. 2007* Sep-Oct;22(5):40, 42-3

Dement JM, Pompeii LA, Østbye T, Epling C, Lipscomb HJ, James T, Jacobs MJ, Jackson G, Thomann W. An integrated comprehensive occupational surveillance system for health care workers. *Am J Ind Med.* 2004;45:528-38.

Hodgson MJ, Reed R, Craig T, Murphy F, Lehmann L, Belton L, Warren N. Violence in healthcare facilities: lessons from the veterans health administration. *J Occup Environ Med.* 2004;46:1158-65.

Hodgson MJ, Mohr D, Drummond D, Bell M, Van Male. Managing disruptive patients in health care: Necessary solutions to a difficult problem, *Amer J Ind Med,* 2012; 55:1009–17. doi:10,1002/ajim22104

Jha AK, Perlin JB, Kizer KW, Dudley RA. Effect of the transformation of the veterans affairs health care system on the quality of care. *N Engl J Med.* 2003;348:2218-27.

Lehmann LS, McCormick RA, Kizer KW. A survey of assaultive behavior in Veterans Health Administration facilities. *Psychiatr Serv.* 1999;50:384-9.

Lehmann LS, Padilla M, Clark S, Loucks S. Training personnel in the prevention and management of violent behavior. *Hosp Community Psychiatry.* 1983;34:40-3.

Mohr DC, Warren N, Hodgson MJ, Drummond DJ. Assault rates and implementation of a workplace violence prevention program in the veterans health care administration. *J Occup Environ Med.* 2011;53:511-6.

Nelson A, Matz M, Chen F, Siddharthan K, Lloyd J, Fragala G. Development and evaluation of a multifaceted ergonomics program to prevent injuries associated with patient handling tasks. *Int J Nurs Stud.* 2006 Aug;43(6):717-33.

Nelson A, Lloyd JD, Menzel N, Gross C. Preventing nursing back injuries: redesigning patient handling tasks. *AAOHN J.* 2003 Mar;51(3):126-34.

Siddharthan K, Nelson A, Weisenborn G. A business case for patient care ergonomic interventions. *Nurs Adm Q.* 2005 Jan-Mar;29(1):63-71.

Siddharthan K, Hodgson M, Rosenberg D, Haiduven D, Nelson A. Under-reporting of work-related musculoskeletal disorders in the veterans administration. *Int J Health Care Qual Assur Inc Leadersh Health Serv.* 2006;19:463-76.

Linking Workers' Compensation Data and Earnings Data to Estimate the Economic Consequences of Workplace Injuries

Seth A. Seabury
RAND Corporation

Introduction

There are many costs associated with workplace injuries and illnesses. Among the most important are the economic losses suffered by injured workers in the form of lost earnings.[1] These losses derive both from uncompensated time out of work during recovery and from residual long-term disability that lowers both the likelihood of working in the future and wages for those who do work. Quantifying earnings losses is important for understanding the magnitude of the problem caused by poor health and safety in the workplace. Earnings loss estimates also provide a useful metric for evaluating the performance of workers' compensation in meeting key policy objectives.

Accurately measuring the economic outcomes for injured workers poses several empirical challenges. Past work has made strides estimating earnings losses by combining data from state workers' compensation systems with other administrative databases on earnings. The purpose of this article is to describe this method, highlight some key lessons learned and identify some areas where more effort is needed.

Overview of the Methodological Approach
Simply put, earnings losses are just the difference between injured workers' expected earnings and their actual earnings after an injury. However, measuring this requires knowledge of a counterfactual: What would injured workers have earned in the absence of an injury? Since what would have happened is fundamentally unknowable, estimating earnings losses requires estimating the uninjured earnings of injured workers.

Others have summarized the development of earnings loss estimates in more detail,[2] so here I provide only a brief summary. The use of workers' compensation data linked to earnings to estimate losses dates back to the 1960s. Early studies took the pre-injury earnings of injured workers and projected expected earnings using aggregate trends in earnings.[3-6] This approach is limited by the assumption that average earnings growth for injured workers mirrors aggregate trends. If injured workers are a nonrandom sample of workers—say, if they had lower expected wage growth—it could introduce bias.

In the late 1990s, researchers began using more refined estimates of expected earnings. The breakthrough was the introduction of a quasi-experimental design: compare the outcomes (in this case, earnings) of the "treated" subjects (injured workers) to a sample of "untreated" control subjects before and after treatment (the date of injury). As long as control workers are selected such that their expected earnings in the post-injury period equal the expected earnings of injured workers, this method produces unbiased estimates of earnings losses.

Past studies have mostly used one of two criteria to identify control workers. The first is to use workers who were injured but with minimal severity and little time out of work (e.g., medical only injuries).[7-9] The difference between injured and "uninjured" worker earnings is estimated controlling for other confounders using multivariate regression. The other commonly used approach is to match injured workers to workers who were never injured, but who worked at the same firm and had very similar earnings to the injured workers prior to the injury.[10-12]

Either approach can provide unbiased estimates of earnings losses, but they hinge on different assumptions. Studies using minor injuries as controls rely on the comparatively minor physical harm suffered to assume that earnings losses from the injury are only experienced in the short-term, with no residual and lasting effects. Studies using uninjured workers as controls assume that closeness in pre-injury earnings at the same firm accurately predicts closeness in post-injury earnings. However, data on workers' compensation typically provides more detailed information on demographic characteristics (e.g., gender) than data on earnings. This means that studies using only data on workers' compensation claimants have better ability to control for confounding factors that could affect labor market outcomes, but at the cost of assuming zero intermediate and long-term earnings losses for minor injuries. Recent evidence is mixed on whether the results are affected by different matching techniques.[12, 13]

Past Findings
Figure 1 gives an example of estimated earnings loss using data from a recent RAND study of permanently disabled workers' compensation claimants in California.[14] The data on workers' compensation claims were from the California Disability Evaluation Unit (DEU) for claims with injury dates from 2000 to 2007. Workers' compensation records were linked to quarterly earnings data for up to 12 quarters prior to the quarter of date of injuries and up to 20 quarters after the date of injury. This estimate used the second method described above: that is, control workers were selected based on having very similar earnings at the same firm as the uninjured workers they were matched to.

The pattern in Figure 1 shows what has been found in most earnings loss studies, particularly for permanently disabled workers: that is, losses are immediate, severe and persistent over time. The earnings of injured and control workers track each other closely in the time leading up to the date of injury. However, beginning in the quarter of injury there is a sharp drop in earnings of about 25-30%. This decline persists over the full 5 years following the date of injury.

While the estimated size of losses from any particular study differs according to the characteristics of the sample used, this overall pattern has been consistently found in most earnings loss studies. Prior to injury, the earnings of injured and control workers follow a similar trend (mitigating though not eliminating concerns of selection on unobserved characteristics, which could bias earnings loss estimates). Some studies have found greater levels of recovery in earnings than witnessed for workers in Figure 1,[9, 11] but most studies still find a strong residual earnings loss even several years after the date of injury.

One of the key findings from this work has been the questionable long-term adequacy of workers' compensation benefits.[15, 16] A study of permanent disability benefits in 5 states showed that workers' compensation benefits replaced less than 50% of pre-tax earnings (in one states as low as 30%) 10 years after an injury.[17] However, this finding does appear to depend on the particular jurisdiction being studied. A study of workers' compensation benefit adequacy in Canada suggested much higher levels of income replacement.[18]

An important feature of earnings loss estimates is that they can be used as outcome variables to identify how economic outcomes for injured workers are affected by the injured workers' own characteristics, policy interventions or other system features. Past studies have used earnings loss estimates to test for gender discrimination among disabled workers,[9] evaluate how workers' compensation reforms affect return to work,[14] and evaluate the accuracy and fairness of disability evaluation systems used to determine compensation.[18-21]

Directions for Future Work
While much has been learned about the short-term and long-term economic consequences of workplace injuries using earnings loss studies, serious knowledge gaps remain. Perhaps the

biggest limitation of the existing literature is a lack of generalizability. Workers' compensation is made up of individual state systems with no centralized, national database of claims. Thus, earnings loss studies have been conducted on a jurisdiction-by-jurisdiction basis. To date there have only been studies done in a handful of US states and Canadian provinces (to my knowledge just California, Minnesota, Michigan, New Mexico, Oregon, Washington, Wisconsin, Ontario and British Columbia). Given the wide diversity in labor market conditions and workforce characteristics across the US, it is possible that outcomes for injured workers could differ significantly. More general conclusions about benefit adequacy could be drawn if we had a wider set of states across which to compare outcomes and income replacement.

A related problem is a lack of consistent earnings loss estimates over long time periods covering different aspects of the business cycle. One study found that local economic conditions within California did affect earnings losses, but there is little evidence about the effect of broader trends in outcomes.[22] To better understand the impact of local and national economic conditions on earnings losses, a larger and more comprehensive database would be needed than has previously been used.

Data limitations have also prevented the study of how earnings losses vary over important individual and employer characteristics that could be important drivers of economic outcomes. Relatively little is known about how losses differ according to different types of jobs because occupation is often not recorded (or recorded inconsistently) in administrative databases. Moreover, age, race and education are all individual characteristics that could affect earnings losses, but a lack of reliable data for both injured workers and controls has prevented their systematic study. The issue of age is particularly important topic of study, because this is often used as a basis to adjust permanent disability benefits, though inconsistently across states.

There is evidence that employer characteristics affect economic outcomes (e.g., workers at larger employers have lower losses).[9, 23, 24] However, it is unknown whether these differences are because of the behavior of employers or due to systematic differences in their workers. In particular, more work is needed to better understand how injured workers are affected by employer accommodations and disability management strategies.

Finally, there are key aspects of workers' compensation systems that could influence economic outcomes that have not been evaluated. The availability, cost and quality of medical treatment for injured workers are all items of intense concern, but it is largely unknown how medical care affects earnings losses in the long term. Similarly, there are legal aspects of the workers' compensation system that could affect outcomes for injured workers, such as litigation that arises over benefit disputes or incentives for employers to adopt disability management programs or worksite accommodations (because of experience rating of insurance premiums, for example). Linking earnings loss estimates to medical treatment and different dispute resolution outcomes could substantially increase our understanding of how these and other factors impact injured workers.

Discussion

When used properly, earnings loss estimates provide an objective measure of a key aspect of the economic burden of workplace injuries and illnesses. This can be useful for evaluating the cost-effectiveness of different safety interventions. Moreover, earnings loss estimates can provide a key metric for assessing how system features, reforms, and policy interventions affect outcomes injured workers. That said, current data limitations make it difficult to exploit these opportunities to their full potential.

A national (or international) sample of earnings loss estimates for injured workers would provide a more robust and general picture of the experience of injured workers. It would also allow researchers to study new topics and take advantage of the different features of individual state policies in a natural experimental framework. Significant progress along these lines could be made by pooling data from the different state programs, or by combining administrative data on earnings with privately held data on workers' compensation claims (e.g., by a large, national insurer). Alternatively, new questions could be added to existing national surveys to better identify injured workers. Exploring these different options should be a priority for future work estimating the economic consequences of workplace injuries and illnesses.

Figure 1. Example of Earnings Loss Estimates Using Permanently Disabled Workers from California

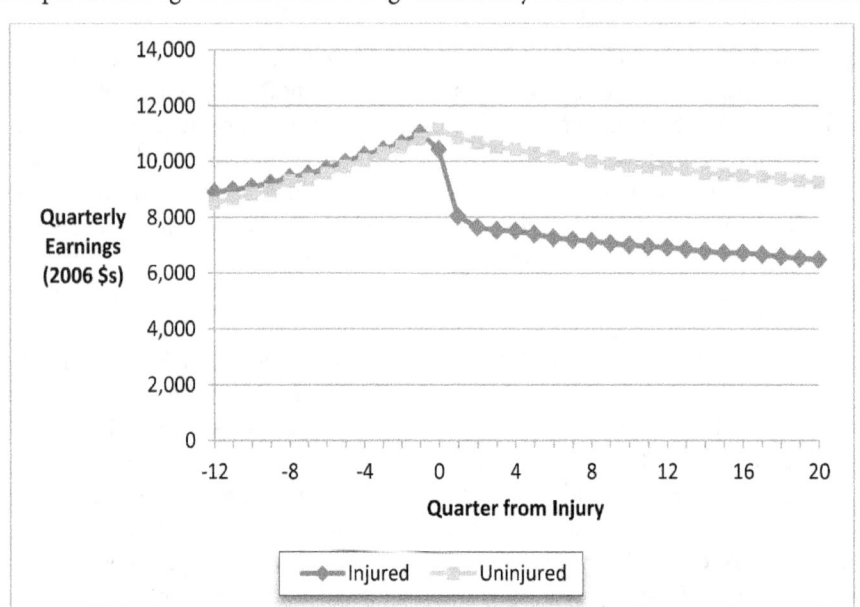

Source: Author's Calculations

---

[1] The development of the two different methods was influenced at least in part by differences in the availability of data for different studies.

[2] The availability of data differed based on the length of time between date of injury and the date of data extraction in early 2010.

[3] More detail on the exact data and methods are available in the study report.

[4] Note that there is a decline in earnings for both the control workers and injured workers. The reason for this is that all workers were required to be working in the quarter of injury (a necessary restriction to identify the at-injury firm). The decline in average control worker earnings after the quarter of injury reflect the natural attrition from the labor market that occurs over time (people retire, get fired, move out of state, etc.)

[5] The National Longitudinal Survey of Youth (NLSY) has been used to study earnings losses,[13] but the geographic information in the NLSY is restricted and the sample size of injured workers could make it hard to identify significant regional or time trends.

[6] For example, California increases permanent partial disability benefits for older workers while Colorado increases them for younger workers.

References

1. Leigh, J., Economic Burden of Occupational Injury and Illness in the United States. *Milbank Quarterly*, 2011. 89(4): p. 728-772.

2. Weil, D., Valuing the economic consequences of work injury and illness: a comparison of methods and findings. *American journal of industrial medicine*, 2001. 40(4): p. 418-437.

3. Cheit, E.F., *Injury and Recovery in the Course of Employment*, 1961: Wiley.

4. Johnson, W.G., P.R. Cullinan, and W.P. Curington, The adequacy of workers' compensation benefits. Research Report of the Interdepartmental Workers' Compensation Task Force, 1978. 6: p. 95-121.

5. Ginnold, R., A follow-up study of permanent disability cases under Wisconsin workers' compensation. Research Report of the Interdepartmental Workers' Compensation Task Force, 1979. 6.

6. Berkowitz, M. and J.F. Burton, Permanent disability benefits in workers' compensation, 1987, Kalamazoo, MI: W.E. Upjohn Institute for Employment Research.

7. Biddle, J., Estimation and analysis of long term wage losses and wage replacement rates of Washington State workers' compensation claimants, Appendix F, in Performance audit of the Washington State workers' compensation system, E. Welch, Editor 1998, Joint Legislative Audit and Review Committee: Olympia, WA.

8. Boden, L.I. and M. Galizzi, Ecnomic consequences of workplace injuries and illnesses: Lost earnings and benefit adequacy. *American Journal of Industrial Medicine*, 1999. 36(5): p. 487-503.

9. Boden, L.I. and M. Galizzi, Income losses of women and men injured at work. *Journal of Human Resources*, 2003. 38(3): p. 722.

10. Peterson, M.A., et al., Compensating permanent workplace injuries : a study of the California system, 1998, Santa Monica, CA: RAND. xvi, 228.

11. Reville, R.T., The Impact of a Permanently Disabling Workplace Injury on Labor Force Participation and Earnings, in The Creation and Analysis of Linked Employer-Employee Data: Contributions to Economic Analysis, J. Haltiwanger, and Haltiwanger, Julia Lane, Editor 1999, Elsevier Science, North-Holland: Amsterdam, London, and New York.

12. Crichton, S., S. Stillman, and D. Hyslop, Returning to Work from Injury: Longitudinal Evidence on Employment and Earnings. *Industrial & Labor Relations Review*, 2011. 64(4): p. 7.

13. Woock, C., Earnings Losses of Injured Men: Reported and Unreported Injuries. *Industrial Relations: A Journal of Economy and Society*, 2009. 48(4): p. 610-628.

14. Seabury, S.A., et al., Workers' Compensation Reform and Return to Work, 2011, Santa Monica, CA: RAND Corporation.

15. Hunt, H.A., Benefit adequacy in state workers' compensation programs. *Social Security Bulletin*, 2004. 65(4): p. 24.

16. Hunt, H.A., Adequacy of earnings replacement in workers' compensation programs: a report of the Study Panel on Benefit Adequacy of the Workers' Compensation Steering Committee, National Academy of Social Insurance, 2004: WE Upjohn Institute.

17. Reville, R.T., et al., An evaluation of New Mexico Workers' Compensation Permanent Partial Disability and return to work, 2001, Santa Monica, CA: RAND. xxvii, 90.

18. Tompa, E., et al., Comparative benefits adequacy and equity of three Canadian workers' compensation programs for long-term disability, 2011, Institute for Work and Health Toronto, ON.

19. Reville, R.T., et al., Comparing severity of impairment for different permanent upper extremity musculoskeletal injuries. *Journal of Occupational Rehabilitation*, 2002. 12(3): p. 205-221.

20. Reville, R.T., et al., An evaluation of California's permanent disability rating system, 2005, Santa Monica, CA: RAND Corp. xxxii, 138.

21. Bhattacharya, J., et al., Evaluating Permanent Disability Ratings Using Empirical Data on Earnings Losses. *Journal of Risk and Insurance*, 2010. 77(1): p. 231-260.

22. Reville, R.T., et al., Trends in earnings loss from disabling workplace injuries in California: the role of economic conditions, 2002, Santa Monica, CA: RAND. xxi, 46.

23. Reville, R.T. and Rand Corporation., Permanent disability at private, self-insured firms a study of earnings loss, replacement, and return to work for workers' compensation claimants, 2000, RAND.

24. Seabury, S.A., et al., Workers' Compensation Experience Rating and Return to Work. *Policy and Practice in Health and Safety*, 2012. 10(1): p. 97-116.

Workers' Compensation Costs in Wholesale and Retail Trade Sectors[1]

Anasua Bhattacharya, Paul Schulte, and Vern Anderson
National Institute for Occupational Safety and Health

Introduction

The wholesale and retail trade (WRT) sector employs nearly 20 million workers. The wholesale trade sector is identified by the Bureau of Labor Statistics' (BLS) North American Industry Classification System (NAICS) code 42, and the retail trade sector is identified by the NAICS codes 44 and 45. According to the Current Population Survey (CPS), wholesale trade sector employment in 2010 was 3.8 million and retail trade sector employment was 15.9 million. About 55 percent of WRT workers were male [BLS 2011a]. In the same year, the WRT sector had 633,500 nonfatal injuries [BLS 2011b] and 502 fatalities [BLS 2012a]. The incidence rate of nonfatal injuries in the wholesale trade sector was 3.3 per 100 full-time equivalent workers, and in the retail trade sector the rate was 4.0 per 100 full-time equivalent workers in 2010. These figures compare to 3.6 per 100 full-time equivalent workers in all private sectors in 2010 [BLS 2012b]. The incidence rate of fatal injuries in the wholesale trade sector was 4.9 per 100,000 full-time equivalent workers, and in the retail trade sector 2.2 per 100,000 full-time equivalent workers. These figures compare to 3.8 per 100,000 full-time equivalent workers in all private sectors in 2010 [http://www.bls.gov/iif/oshwc/cfoi/cfoi_revised10.pdf]. The incidence rates for fatality in wholesale trade and nonfatality in retail trade are higher than the average of all private industries. Studies have shown that at the 4- and 5- digit NAICS codes of WRT industries, a wide range of work activities and physical hazards may cause a substantial risk [NIOSH 2006]. These workplace hazards cause fatal and nonfatal injuries that result in an immense loss to the employers, employees, and the economy. Some of these losses are covered by the Workers' Compensation (WC) system, and the rest are distributed to the employers in the form of lost productivity, to the employees and their family members as pain and suffering, and to society [Safe Work Australia 2012]. This study focuses on the indemnity costs and medical costs of fatal and nonfatal injuries in WRT for the years 2003 through 2007. WC costs are used to estimate the losses in WRT sectors by body parts injured and nature of injury.

Data

Primary data for this study are obtained from BLS and the National Council on Compensation Insurance (NCCI). BLS provides the number of fatal and nonfatal injuries by the nature of injury and body parts injured. The number of fatalities is obtained from Census of Fatal Occupational Injuries (CFOI) research files. WC data on indemnity costs (WC payments for lost wages) and medical costs are obtained from NCCI by the nature of injury and body parts injured. The NCCI data has about 1.4 million claims on WRT for the years 2003 through 2007. The NCCI WC costs utilized are incurred costs and not current paid costs. Incurred costs are forward-looking, that is, the amount that needs to be set aside today to account for current and any future costs [Leigh and Marcin 2012].

---

[1] Disclaimers: The findings and conclusions in this report are those of the authors and do not necessarily represent the views of the National Institute for Occupational Safety and Health. This research was conducted with restricted access to Bureau of Labor Statistics (BLS) data. The views expressed here do not necessarily reflect the views of the BLS.

Methods

The medical cost per claim and the indemnity cost per claim obtained from NCCI data are used as the average medical cost and the average indemnity cost. These average medical costs and average indemnity costs are classified by the different body parts injured and the nature of the injury. Total medical costs are obtained from the product of average medical cost from NCCI and number of nonfatal injuries and fatal injuries from BLS. Total indemnity costs are estimated from the product of average indemnity cost from NCCI and number of nonfatal injuries and fatal injuries from BLS. Total WC costs are estimated as the sum of total medical costs and total indemnity costs. The number of fatalities by nature of injury for most categories was too small to report, so that is not included in the study.

Results

Table 1 shows the number of nonfatal injuries, average WC costs in 2010 dollar values (average cost is the sum of average medical cost and average indemnity cost), total WC costs (TWC) and the percentages of injuries by different body parts injured from 2003 through 2007. Both the number of nonfatal injuries and average WC costs decreased over the years (except for the number of arm injuries and the average WC costs of neck injuries). The results show that from the year 2003 through 2007, the WRT sector had a decrease of 8% for "All" nonfatal injuries, a decrease of 34% for "All" average WC (AWC) costs and a decrease of 39% for "All" TWC costs. The frequencies and percentages of back injuries (59,194 and 24% in 2003, and 48,190 and 22% in 2007) are highest among all the different types of body parts injured, followed by multiple body parts and trunk injuries. The AWC costs are highest for the neck injuries ($36,448 in 2003, and $37,711 in 2007) followed by the shoulder injuries ($29,161 in 2003, and $22,306 in 2007). The TWC costs are highest for back injuries ($1.5 billion in 2003, and $0.7 billion in 2007).

Table 2 shows the number of nonfatal injuries, AWC and TWC costs in 2010 dollar values, and percentages of injuries by the different nature of injuries from 2003 through 2007. The results are similar to Table 1, suggesting that the number of nonfatal injuries and AWC costs by different nature of injuries decreased over the years (except for number of amputations and number of fractures). The frequencies and percentages of sprain and strain (108,537 and 45% in 2003, and 89,008 and 40% in 2007) are highest among all the different types of nature of injuries, followed by contusion and concussion. AWC costs are the highest for amputations ($52,566 in 2003, and $43,505 in 2007) followed by fractures ($27,548 in 2003, and $22,809 in 2007). The TWC costs are highest for sprain and strain ($2.1 billion in 2003, and $1.1 billion in 2007).

Table 3 shows the number of fatal injuries, AWC and TWC costs in 2010 dollar values and percentages of the injuries by the different body parts injured from 2003 through 2007. The number of fatal injuries increased from 2003 (545) to 2005 (613) and then decreased to 551 in 2007. AWC costs decreased from $334,537 in 2003 to $212,030 in 2004 and then increased to $280,915 in 2007 for all fatalities. Frequencies and percentages of multiple body parts injured (178 and 33% in 2003, and 232 and 42% in 2007) and head injuries (165 and 30% in 2003, and 122 and 22% in 2007) are the highest among all the body parts injured. The AWC varied a lot during this period for the different body parts injured. AWC costs for neck injuries were the highest for 2007 ($550,711) while AWC costs for head injuries were the highest for 2003 ($411,496). The TWC costs are highest for 2003 head injuries ($68 million) and 2007 multiple body parts injuries ($68 million).

Table 4 shows the medical costs, indemnity costs, and total costs for fatal and nonfatal injuries separately and together. It also shows the total medical costs and indemnity costs in 2010 dollar values for all injuries and the total estimated WC costs for the years 2003 through 2007. The results suggest that the medical costs decreased for both fatal (from $61 million in 2003 to $10 million in 2007) and nonfatal injuries (from $1.1 billion in 2003, to $0.9 billion in 2007), but the indemnity costs increased for fatal injuries (from $121 million in 2003 to $145 million in 2007) and decreased for nonfatal injuries (from $3.7 billion in 2003 to $2.1 billion in 2007). The estimated total WC costs (sum of medical costs and indemnity costs) for all fatal and nonfatal injuries in the WRT sector decreased from $4.9 billion in 2003 to $3.1 billion in 2007, a decrease in 38 percent. Chart 1 shows that the total WC costs, total indemnity costs and total medical costs have dropped similarly in these years. It also shows that total non-fatal WC costs and total fatal WC costs have almost remained the same during these years.

Discussion

Many studies have demonstrated that WC systems do not compensate for all fatal and nonfatal injuries, as there are conditions and incentives that discourage the submission of a WC claim, and the compensation itself is inadequate [Azaroff et al. 2002; Leigh and Robbins 2004; Bonauto et al. 2010]. Previous studies focusing on the WRT sector have concluded that the health burden of occupational injuries and fatalities is substantial for the WRT sector [Anderson et al. 2010]. The value of determining the true economic burden of occupational injuries and illnesses lies in the potential benefit for the employers, employees, and society from reducing the hazards and improving workplace safety. This is the first attempt to estimate the medical costs and indemnity costs of fatal and nonfatal injuries for the WRT sector. Due to the large number of employees in this sector, even a small increase in injury rates can significantly affect the burden for the employers, employees, and society. The outcomes obtained suggest that the estimated total WC costs have decreased from 2003 through 2007, yet they remain high. This decline is due to the drop in the number of nonfatal injuries and average WC costs. A reason behind this decline in WC costs for nonfatal injuries could be increased under-reporting over the years. According to BLS data [BLS 2004, 2005, 2006, 2007, 2008], disabling injuries (that is, injuries that involve days away from work and cases of job transfer restriction due to injuries) are approximately 55 percent of all injuries. Nondisabling injuries (that is, injuries that do not require days away from work) are approximately 45 percent. Leigh et al. [2000] suggested that about 35% of nondisabling injuries are underreported and 20% of disabling injuries are underreported. Therefore, an average of 28% underreporting can be assumed for all nonfatal injuries; with this assumption, the total costs of nonfatal injuries will be $6.9 billion in 2003, and $4.2 billion in 2007 in 2010 dollar values. Another reason for the decline in indemnity costs could be that injured workers are brought back to work earlier and better accommodated while they are on the mend.

Medical costs for fatalities are highly unstable, varying from $9.54 million (in 2007) to $61.38 million (in 2003) for fatal injuries. This can be both because of highly variable numbers of fatalities and because of the strong rightward skew of the cost per case distribution (high cost outliers).

The results obtained by body parts injured and nature of injury (data not shown) suggest that the total costs are highest for back injuries, fractures, and sprain and strains. Many of the employers in the WRT sector are small businesses with low profit margins. Therefore, any workplace injury is more detrimental to these employers compared with large corporations. Controlling exposures triggering these injuries will prevent the injuries, improve productivity, and will reduce losses in the economy.

Limitation and Future Research
This study estimates the medical costs and indemnity costs of fatal and nonfatal injuries in WRT. A true economic burden will incorporate indirect costs of fatal and nonfatal injuries accounting for the pain and suffering of the injured workers and the underreporting of occupational injuries that are not included in this study. There is also a difference between the number of WC claims and BLS counts; the BLS capture rate is smaller than the WC

capture rate [Boden et al. 2010], and this study utilized BLS counts with no adjustments for underreporting. There is a need for more research to determine the factors contributing to the most expensive treatments, such as back injuries and head injuries, which have the highest WC average medical costs. Another needed extension is an analysis incorporating all major industries. Work-related injury data are publicly available from BLS, but the WC data are only available from individual WC bureaus, some of which are so expensive it is impractical to conduct comprehensive research studies. Different states have different WC systems, and they cannot be directly linked to BLS injury data. Improved linkage between the WC data and BLS injury data would help researchers predict the true economic burden of workplace injuries and fatalities.

References
Anderson VP, Schulte PA, Sestito J, Linn H, Nguyen LS [2010]. Occupational fatalities, injuries, illnesses, and related economic loss in the wholesale and retail trade sector. Am J Ind Med 53(7):673–685.

Azaroff LS, Levenstein C, Wegman DH [2002]. Occupational injury and illness surveillance: conceptual filters explain underreporting. Am J Public Health 92(9):1421–1429.

BLS [2004]. Workplace injuries and illnesses—2003. Press release, December 14, 2004. Washington, DC: U.S. Department of Labor, Bureau of Labor Statistics [http://www.bls.gov/iif/oshwc/osh/os/osnr0021.pdf].

BLS [2005]. Workplace injuries and illnesses—2004. Press release, November 17, 2005. Washington, DC: U.S. Department of Labor, Bureau of Labor Statistics [http://www.bls.gov/iif/oshwc/osh/os/osnr0023.pdf].

BLS [2006]. Workplace injuries and illnesses—2005. Press release, October 19, 2006. Washington, DC: U.S. Department of Labor, Bureau of Labor Statistics [http://www.bls.gov/iif/oshwc/osh/os/osnr0025.pdf].

BLS [2007]. Workplace injuries and illnesses—2006. Press release, October 16, 2007. Washington, DC: U.S. Department of Labor, Bureau of Labor Statistics [http://www.bls.gov/iif/oshwc/osh/os/osnr0028.pdf].

BLS [2008]. Workplace injuries and illnesses—2007. Press release, October 23, 2008. Washington, DC: U.S. Department of Labor, Bureau of Labor Statistics [http://www.bls.gov/iif/oshwc/osh/os/osnr0030.pdf].

BLS [2011a]. Household data annual averages. Washington, DC: U.S. Department of Labor, Bureau of Labor Statistics [http://www.bls.gov/cps/cpsa2010.pdf].

BLS [2011b]. Workplace injuries and illnesses—2010. Press release, October 20, 2011. Washington, DC: U.S. Department of Labor, Bureau of Labor Statistics [http.release/archives/osh_10202011.pd p://www.bls.gov/newsf].

BLS [2012a]. Revisions to the 2010 CFOI Counts. Washington, DC: U.S. Department of Labor, Bureau of Labor Statistics [http://www.bls.gov/iif/oshwc/cfoi/cftb0250.pdf].

BLS [2012b]. Revisions to the 2010 census of fatal occupational injuries (CFOI) counts. Washington, DC: U.S. Department of Labor, Bureau of Labor Statistics [http://www.bls.gov/iif/oshwc/cfoi/cfoi_revised10.pdf]. Accessed August 16, 2012.

Boden LI, Nestoriak N, Pierce B [2010]. Using capture-recapture analysis to identify factors associated with differential reporting of workplace injuries and illnesses. Section on Survey Research Methods—JSM 2010 [http://www.bls.gov/osmr/pdf/st100300.pdf].

Bonauto D, Adams D, Smith C, Fan JZ, Silverstein B, Foley M [2010]. Language preference and non-traumatic low back disorders in Washington State workers' compensation, Am J Ind Med 53(2):204–215 [http://onlinelibrary.wiley.com/doi/10.1002/ajim.20740/pdf].

Leigh JP, Marcin JP [2012]. Workers' compensation benefits and shifting costs for occupational injury and illness. *J Occup Environ Med* 54(4):445–450.

Leigh JP, Markowitz SB, Fahs MC, Landrigan P [2000]. Costs of occupational injuries and illnesses. Ann Arbor, MI: University of Michigan Press.

Leigh JP, Robbins JA [2004]. Occupational disease and workers' compensation: coverage, costs and consequences. *The Milbank Q* 82(4):689–721.

NIOSH [2006]. NIOSH program portfolio, wholesale and retail trade, sector description. Cincinnati, OH: U.S. Department of Health and Human Services, Centers for Disease Control and Prevention, National Institute for Occupational Safety and Health, DHHS (NIOSH) [www.cdc.gov/niosh/programs/wrt/sector.html].

Safe Work Australia [2012]. The cost of work-related injury and illness for Australian employers, workers and the community: 2008–09. Canberra, Australia: Safe Work Australia [http://www.safeworkaustralia.gov.au/sites/SWA/AboutSafeWorkAustralia/WhatWeDo/Publications/Documents/660/Cost%20of%20Work-related%20injury%20and%20disease.pdf].

Table 1. Number, Average WC Costs, Total WC Costs (in 2010 dollar values) and Percentages of Injuries by Body Parts Injured for Nonfatal Injuries

Body Parts	Number/Costs	2003	2004	2005	2006	2007
Arm	Number	9,695	9,216	10,101	9,225	10,064
	Percent	3.99	3.85	4.28	4.20	4.51
	AWC Costs ($)	19,119	18,421	17,234	16,091	13,853
	TWC Cost ($ mil.)	185	170	174	148	139
Back	Number	59,194	59,858	53,398	50,338	48,190
	Percent	24.36	24.99	22.63	22.90	21.61
	AWC Costs ($)	26,280	22,785	20,383	16,411	14,096
	TWC Cost ($ mil.)	1,556	1,364	1,088	826	679
Head	Number	15,879	14,054	14,921	15,133	15,547
	Percent	6.53	5.87	6.32	6.88	6.97
	AWC Costs ($)	26,717	27,819	22,027	19,518	18,583
	TWC Cost ($ mil.)	424	391	329	295	289
Multiple Body Parts	Number	21,446	21,197	20,284	19,105	21,372
	Percent	8.82	8.85	8.60	8.69	9.58
	AWC Costs ($)	24,864	25,375	25,380	20,983	18,167
	TWC Cost ($ mil.)	533	538	515	401	388
Neck	Number	3,742	4,285	3,557	3,180	3,345
	Percent	1.54	1.79	1.51	1.45	1.50
	AWC Costs ($)	36,448	33,173	34,129	28,506	37,711
	TWC Cost ($ mil.)	136	142	121	91	126
Shoulder	Number	15,916	15,252	15,280	15,133	14,948
	Percent	6.55	6.37	6.48	6.88	6.70
	AWC Costs ($)	29,161	28,163	26,345	24,111	22,306
	TWC Cost ($ mil.)	464	430	403	365	333
Trunk	Number	16,782	16,414	17,086	14,773	14,386
	Percent	6.90	6.85	7.24	6.72	6.45
	AWC Costs ($)	18,753	17,573	16,067	14,359	12,888
	TWC Cost ($ mil.)	315	288	275	212	185
All	Number	243,045	239,524	235,976	219,802	223,046
	AWC Costs ($)	19,778	18,341	16,892	14,464	13,079
	TWC Cost ($ mil.)	4,807	4,393	3,986	3,179	2,917

The percentages do not add up to 100 percent as the 'Other' category is not included in the table.

Table 2. Number, Average WC Costs, Total WC Costs (in 2010 dollar values) and Percentages of Injuries by Nature of Injuries for Nonfatal Injuries

Nature of Injury	Number/ Costs	2003	2004	2005	2006	2007
Amputation	Number	903	1,327	1,263	903	1,005
	Percent	0.37	0.55	0.54	0.41	0.45
	AWC Costs ($)	52,566	44,392	41,180	42,003	43,505
	TWC Cost ($ mil.)	47	59	52	38	44
Burns	Number	2,894	3,364	3,428	3,226	2,617
	Percent	1.19	1.40	1.45	1.47	1.17
	AWC Costs ($)	10,923	12,650	12,181	14,633	8,271
	TWC Cost ($ mil.)	32	43	42	47	22
Contusion/ Concussion	Number	25,150	24,284	23,538	21,492	21,787
	Percent	10.35	10.14	9.97	9.78	9.77
	AWC Costs ($)	15,814	14,184	12,728	10,700	9,324
	TWC Cost ($ mil.)	398	344	300	230	203
Carpal Tunnel Syndrome	Number	3,290	2,682	2,627	2,064	2,193
	Percent	1.35	1.12	1.11	0.94	0.98
	AWC Costs ($)	25,711	22,441	21,286	21,537	20,228
	TWC Cost ($ mil.)	85	60	56	44	44
Fracture	Number	15,852	16,663	17,547	16,220	17,815
	Percent	6.52	6.96	7.44	7.38	7.99
	AWC Costs ($)	27,548	27,288	24,798	24,858	22,809
	TWC Cost ($ mil.)	437	455	435	403	406
Sprain/ Strain	Number	108,537	106,316	102,979	94,307	89,008
	Percent	44.66	44.39	43.64	42.91	39.91
	AWC Costs ($)	19,943	18,610	17,241	14,692	12,770
	TWC Cost ($ mil.)	2,165	1,979	1,775	1,386	1,137
All	Number	243,045	239,524	235,976	219,802	223,046
	AWC Costs ($)	19,778	18,341	16,892	14,464	13,079
	TWC Cost ($ mil.)	4,807	4,393	3,986	3,179	2,917

The percentages do not add up to 100 percent as the 'Other' category is not included in the table.

Table 3. Number, Average WC Costs, Total WC Costs and Percentages of Injuries by Body Parts Injured for Fatal Injuries (in 2010 dollar values)

Body Parts	Number/Costs	2003	2004	2005	2006	2007
Back	Number	12	11	15	7	12
	Percent	2.20	1.89	2.45	1.20	2.18
	AWC Costs ($)	349,538	188,154	132,078	265,965	0
	TWC Cost ($ mil.)	4	2	2	2	0
Head	Number	165	161	154	159	122
	Percent	30.28	27.66	25.12	27.37	22.14
	AWC Costs ($)	411,496	251,434	161,830	261,122	222,131
	TWC Cost ($ mil.)	68	40	25	42	27
Multiple Body Parts	Number	178	213	230	219	232
	Percent	32.66	36.60	37.52	37.69	42.11
	AWC Costs ($)	309,514	200,479	241,481	259,405	294,167
	TWC Cost ($ mil.)	55	43	56	57	68
Neck	Number	11	14	13	12	20
	Percent	2.02	2.41	2.12	2.07	3.63
	AWC Costs ($)	294,107	243,724	81,993	399,984	550,711
	TWC Cost ($ mil.)	3	3	1	5	11
All	Number	545	582	613	581	551
	AWC Costs ($)	334,537	212,030	223,114	260,655	280,915
	TWC Cost ($ mil.)	182	123	137	151	155

The percentages do not add up to 100 percent as the 'Other' category is not included in the table.

Table 4. Medical Costs and Indemnity Costs of Nonfatal and Fatal Injuries (Mil. $) (in 2010 dollar values)

Year	Nonfatal Injuries			Fatal Injuries			All Injuries		
	Medical ($)	Indemnity ($)	Total ($)	Medical ($)	Indemnity ($)	Total ($)	Medical ($)	Indemnity ($)	Total ($)
2003	1,075	3,732	4,807	61	121	182	1,136	3,853	4,989
2004	1,025	3,368	4,393	13	110	123	1,038	3,478	4,516
2005	985	3,001	3,986	21	116	137	1,007	3,116	4,123
2006	874	2,305	3,179	21	130	151	895	2,435	3,331
2007	865	2,052	2,917	10	145	155	875	2,197	3,072

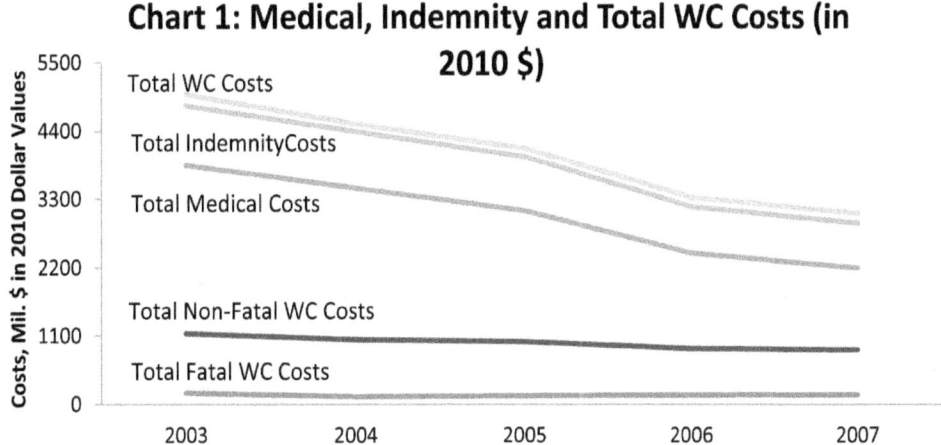

Linking Workers' Compensation and Group Health Insurance Data to Examine the Impact of Occupational Injury on Workers' and their Family Members' Health Care Use and Costs: Two Case Studies[1]

Abay Asfaw, Regina Pana-Cryan, Tim Bushnell, Roger Rosa, Rebecca Mao
National Institute for Occupational Safety and Health

Background

Linking workers' compensation (WC) and group health insurance (GHI) data provides information that allows researchers to follow the pre- and post-work injury health status of both workers and their family members. Although the use of such administrative data for research has some drawbacks, the use of medical and workers' compensation claims data also avoids the limitations that can be associated with surveys, including issues of recall and self-report. The objective of the two case studies described here was to examine the impact of occupational injury on injured workers' and their family members' GHI health care use and costs. In the first case study, we examined the incidence and costs of hospitalization among family members before and after occupational injury. In the second study, we examined GHI utilization and costs following acceptance or denial of WC medical claims.

Case study 1

Incidence and Costs of Family Member Hospitalization Following Injuries of Workers' Compensation Claimants[2]
Abay Asfaw, Regina Pana-Cryan, Tim Bushnell
National Institute for Occupational Safety and Health

Introduction

The objective of this study was to determine whether occupational injuries for which WC claims were filed were associated with subsequent short-term increases in inpatient medical care for family members. There are several reasons why occupational injury might have consequences for the family. First, as indicated by Weil (2001), occupational injuries may reduce family income in two ways, since WC benefits do not fully replace regular wages and family members also might not be able to seek employment or stay as fully employed while caring for an injured worker as they were before the injury. In the most difficult situations, families may be forced to sell their assets, leave or change school, or move (Morse et al. 1998). Second, family members may also have to shoulder greater physical burdens to care for the injured worker and perform household tasks to which the injured worker cannot contribute (Morse et al., 1998; Strunin and Boden, 2004). Third, the psychological distress of the injured worker might also lead to stress and psychological problems among family members (Morse et al., 1998; Strunin and Boden, 2004). As a result of these impacts, families of injured workers may also experience additional health problems. Using data from Canada, Brown et al. (2007) found that medical care use was higher for the families of injured workers over the five year period following the year of injury. In this study, we focused on hospitalizations as indicators of the most severe impacts on health and medical care use and cost of family members. We also

---

[1] The findings and conclusions in this report are those of the authors and do not necessarily represent the views of the National Institute for Occupational Safety and Health.
[2] The full paper, upon which this discussion is based, has been published: Asfaw, A., Pana-Cryan, R. and Bushnell, P. T. (2012), Incidence and costs of family member hospitalization following injuries of workers' compensation claimants. Am. J. Ind. Med., 55: 1028–1036. doi: 10.1002/ajim.22110

focused on short periods of time (3 months) before and after injury. We hypothesized that occupational injury would increase the incidence and costs of hospitalization among workers' families, and that the impact would be higher for family members of more severely injured (SI) workers.

Data and Method

We used the MarketScan Health and Productivity Management (HPM) and Commercial Claims and Encounters (CCE) databases compiled by Thomson Reuters. The data contain information on WC and GHI claims of injured workers' and family members, respectively. Eighteen employers (all clients of Thomson Reuters) provided employee data for the HPM database. The WC claims information in HPM includes an enrollment id, the date of injury, the status of claims (closed /open /reopened), and the amount of indemnity and medical payments. We used the HPM database to identify workers who suffered an occupational injury between 2002 and 2005, and whose WC claim was closed by December 31, 2006 (the last date of data availability at the time of our analysis). An occupational injury was classified as severe if the injured worker received indemnity payments and stayed away from work for at least seven working days following injury. The CCE database includes data files for inpatient, outpatient, and pharmacy GHI claims for workers and their family members. The claims information in CCE includes enrolment id, dates of service, diagnoses, procedures, and payments. Hospitalization data for family members of injured workers were extracted from the CCE inpatient data files for the period between January 1, 2002 and December 31, 2005. We linked the HPM and the CCE files using the anonymous and unique 'enrollment id' variable. We used a conditional logistic regression to estimate the odds ratio of family hospitalization three months before and after occupational injury.

We chose to focus on comparison of 3-month periods before and after injury for two reasons. First, we found that the incidence rate of family hospitalizations rose over the first three months following occupational injury and then fell to approximately the pre-injury rate in the sixth month, so that comparison of 3-month periods might increase the likelihood that differences of statistical significance are detectable. Second, a rise in hospitalization rates within a short time after injury is more plausibly linked to the injury and would be virtually unaffected by long term trends. To observe the GHI medical claims of family members within the three months before and after occupational injury, workers injured before April 1, 2002 and after September 30, 2005 were excluded from the analysis. Before-after comparisons were carried out separately for the families of SI workers and the families of all injured workers. These before-after comparisons addressed our hypothesis that incidence and costs of family hospitalization would be higher following occupational injury.

Results

We used a before-after analysis to compare the odds and costs of family hospitalization three months before and after occupational injury for 18,411 families, 15.7% of whom were SI. Since the claims of each family were observed twice (three months before and three months after occupational injury), the total sample size was 36,822 observations. Table 1 presents the conditional logistic regression results, with odds ratios for family hospitalization after injury versus before injury, and 95% confidence intervals (CI).

Among families of all injured workers, in the three months following occupational injury, the odds of at least one family member being hospitalized were 31% higher than in the three months preceding injury. Among the families of SI workers, the odds of hospitalization were 56% higher in the three months following injury. Because there was no evidence of change in the cost per hospitalization, hospitalization costs were estimated to have increased by approximately the same percentage as the odds of hospitalization.

These results support our hypotheses but should be interpreted with caution for several reasons, including the following. First, it may be possible that the work injury could alter family decisions about undergoing hospitalization, although we could not identify a clear reason that this is responsible for our results. Second, we did not include data on health care services that were not directly attributable to a stay in the hospital or for which claims were not filed. Third, the 3-month comparison periods were designed to capture only short run impacts of occupational injury. Fourth, costs may also have been underestimated due to exclusion of WC cases that were not closed by December 31, 2006. If WC cases of more severe injuries take longer to close, this could have reduced the number and average severity of SI workers in our data set. Finally, the findings may not generalize to segments of the U.S. working population that were underrepresented in the data set we used.

Conclusion

The impact of occupational injury may extend beyond the workplace and adversely affect the health and inpatient care use of family members. To further explore the complex pathways between an occupational injury and the health of family members, future research could focus on the specific nature of occupational injuries (e.g. acute versus cumulative trauma) associated with increases in family health problems, as well as the specific nature of these problems.

<u>Case Study 2</u>

Group Health Insurance Utilization and Cost Following Acceptance or Denial of Workers Compensation Medical Claims
Abay Asfaw, Roger Rosa, Rebecca Mao
National Institute for Occupational Safety and Health

Introduction

Occupational injuries impose high costs on the U.S. healthcare system. Evidence also suggests that workers with known or suspected occupational injuries and illnesses may not file for WC benefits due to fear of disciplinary action, stigmatization, harassment, or denial of benefits. (Biddle et al., 1998; Conway and Svenson, 1998; Rosenman et al., 2000; Morse et al., 2000). Even if some workers apply for WC benefits, employers could dispute the work-relatedness of an injury or condition or challenge its severity. As a result, WC claimants might not receive full indemnity or medical payments or their claim could be totally denied (Ellenberger, 2000; Dembe, 2001; Boden et al., 2001). Based on data from the 2007 Behavioral Risk Factor Surveillance System (BRFSS), CDC (2010) indicated that successful WC claims for medical costs ranged from 47% in Texas to 77% in Kentucky. Leigh & Robbins (2012) indicated that the WC system does not adequately cover the costs of occupational injuries and illnesses, resulting in workers use of other insurance programs to help pay for those costs. Using macro level data from the Bureau of Labor Statistics (BLS) and the National Council on Compensation Insurance (NCCI) and the total costs of occupational injuries and illnesses estimated from Leigh (2011), Leigh and Marcin reported that for medical costs not covered by WC, other insurance programs covered $14,22 billion, Medicare covered $7.16 billion and Medicaid covered $5.47 billion. This study complements such macro level studies by estimating GHI utilization and cost differences between workers whose WC medical claims were accepted and denied using individual level WC and GHI utilization information within a short period after the incidence of occupational injury.

Data and method

The 2002-2005 Thomson Reuters MarketScan Health and Productivity Management (HPM) and Commercial Claims and Encounter (CCE) data described above were used. Overall 52,046 workers who were injured and filed for WC benefits between 2002 and 2005 were used for analysis. Workplace injury was defined by filing for WC indemnity and medical benefits and a WC medical claim was considered denied if no medical costs were paid from the WC program. GHI utilization and costs were measured using outpatient and inpatient GHI records within two weeks before and after the occurrence of an occupational injury. Two-week pre- and post-injury periods were chosen to reduce the influence of other unobservable factors that might affect the health status of the injured workers. Utilization was defined as at least one outpatient or inpatient visit during the time under consideration. Costs were determined separately for the two weeks before and after the occupational injury as the total amount of money paid by GHI during each two-week period.

Results

Overall, 17% and 1% of injured workers used outpatient and inpatient GHI during the study period, respectively. In the two weeks before an occupational injury, 18.8% of workers whose WC medical claims were accepted and 19.9% of workers whose WC medical claims were denied used outpatient GHI at least once. Within two weeks following an occupational injury, GHI utilization for outpatient services increased to 30.4% and 37.8% for workers whose claims were accepted and denied, respectively. Inpatient GHI utilization also increased from 0.05 to 0.1% and from 0.31 to 0.97% for injured workers whose claims were accepted and denied, respectively. All of these differences were statistically significant.

We used logistic regression to examine outpatient and inpatient utilization of group health insurance within two weeks after injury while controlling for pre-injury utilization and other factors. Covariates included in the model were pre-injury health-care utilization, sex, age, hourly versus salaried compensation, union membership status, health plan type, industry, and region of WC claimants. Separate regression equations were estimated for outpatient and inpatient services. The results are presented in Table 2. Holding all other factors constant, the odds of WC claimants whose medical claims were denied using GHI outpatient services at least once within two weeks after injury was 30% higher than that of WC claimants whose medical claims were accepted. The effect was much stronger in the case of GHI inpatient service utilization.

We also estimated the effect WC medical claims denial on the unconditional outpatient and inpatient GHI costs, and part of the results are presented in Figure 1. Denial of WC claims increased outpatient and inpatient GHI costs by 45% and 239%, respectively, controlling for all covariates included in the model.

To give the issue a national perspective, we extrapolated our cost estimates following WC claim denials to national injury figures provided by the Bureau of Labor Statistics (BLS). According to BLS, more than 5 million nonfatal occupational injuries and illnesses were reported per year during our study period. Based on a WC claim rejection range of 19.4% in our sample to 39% in a CDC report (CDC, 2010), denial of WC medical claims could cost other parts of the health care system between $245 to $484 million within two weeks after injury.

The study has the following limitations. First, we did not have any information about why the medical claims were denied. If most of denied claims were not work-related, our results could overestimate the impact of WC denial on the GHI. Second, we did not consider workers who were injured but did not apply for WC. Third, to reduce the effect of other unobservable factors that might affect the health status of the injured workers, we considered costs incurred only within two weeks before and after injury. Costs incurred after two weeks of injury could be substantial. Finally, the data we used were restricted to large employers who were clients of Thomson Reuters and all of the workers had GHI. This might not represent the U.S. working population.

References

Biddle J, Roberts K, Rosenman KD, Welch EM. [1998]. What percentage of workers with work-related illnesses receive workers' compensation benefits? *Journal of Occupational and Environmental Medicine* 40(4): 325-331.

---

[3] In the unconditional analysis we considered all WC claimants irrespective of their GHI utilization.

Boden LI, Biddle EA, Spieler EA. [2001]. Social and economic impacts of workplace illness and injury: current and future directions for research. *Am J Indust Med* 40:398–402.

Brown JA, Shannon HS, McDonough P, Mustard CP. [2007]. Healthcare use of families of injured workers before and after a workplace injury in British Columbia, Canada. *Healthcare Policy* 2(3):e121-e129.

Centers for Disease Control and Prevention (CDC). [2010]. Proportion of workers who were work-injured and payment by workers' compensation systems--10 states, 2007. *Morbidity and Mortality Weekly Report* 59(29):897-900.

Conway H, Svenson J. [1998]. Occupational injury and illness rates, 1992–96: why they fell. *Monthly Labor Rev* November:36–58.

Dembe AE. [2001]. Access to medical care for occupational disorders: difficulties and disparities. *J Health Soc Policy* 12(4):19–33.

Ellenberger JN. [2000]. The battle over workers' compensation. *New Solutions*. 10:217–236.

Leigh JP, Marcin PJ. [2012]. Workers' Compensation Benefits and Shifting Costs for Occupational Injury and Illness. *Journal of Occupational & Environmental Medicine* 54(4):445-450.

Leigh JP, Robbins JA. [2004]. Occupational disease and workers' compensation, costs, and consequences. *Milbank Quarterly* 82:689–721.

Morse T, Dillon C, Warren N, Levenstein C, Warren A. [1998]. The economic and social consequences of work-related musculoskeletal disorders: The Connecticut upper-extremity surveillance project. *Int. J Occup Environ Health* 4:209-216.

Morse T, Dillon C, Warren N. [2000]. Reporting of work-related musculoskeletal disorder (MSD) to workers' compensation. *New Solutions* 10:281–292.

Rosenman KD, Gardiner JC, Wang J, Biddle J, Hogan A, Reilly MJ, Roberts K, Welch E. [2000]. Why most workers with occupational repetitive trauma do not file for workers' compensation. *Journal of Occupational and Environmental Medicine* 42(1):2-34.

Strunin L, Boden L. [2004]. Family consequences of chronic back pain. *Soc. Sci. Med* 58:1385–1393.

Weil D. [2001].Valuing the economic consequences of work injury and illness: A comparison of methods and findings. *Am. J Ind. Med* 40:418–437.

Table 1. Conditional logistic regression results: odds of one or more family hospitalizations three months after versus three months before occupational injury

	All injured workers	Severely injured workers
Odds ratio	1.31	1.56
Z-score	3.17	2.18
P>|z|	0.002	0.029
95% Confidence Interval	1.11 - 1.55	1.05 - 2.34
Number of observations[†]	1,340	212

† The conditional logistic regression analysis procedure employs only observations for families with a change in hospitalization status before and after injury.

Table 2. Determinants of group health insurance utilization within two weeks after injury: Logistic regression results

Variables	Utilization of group health insurance			
	Outpatient		Inpatient	
	OR	95% CI	OR	95% CI
WC medical claim status (1 if denied & 0 otherwise)	1.295***	1.233 - 1.361	3.340***	2.511 - 4.441
Outpatient visit 15 days before injury§	3.332***	3.181 - 3.491		
Male	0.762***	0.728 - 0.798	1.191	0.895 - 1.585
Age	1.014***	1.012 - 1.016	1.027***	1.012 - 1.041
Paid hourly (1 if yes & 0 otherwise)	0.808***	0.750 - 0.871	0.754	0.457 - 1.243
Member of a union	0.921***	0.874 - 0.971	0.89	0.611 - 1.298
Region†				
Industry†				
Health plan type†				
Observations	51990		51859	
Wald chi² (Prob > chi²)	3847 (0.001)		151 (0.001)	
Pseudo R2	0.06		0.04	
Log pseudolikelihood	-30472.727		-1403.5365	

*** p<0.01, ** p<0.05, * p<0.1
§ In the inpatient equation the variable was dropped due to perfect collinearity.
† Results omitted for brevity.

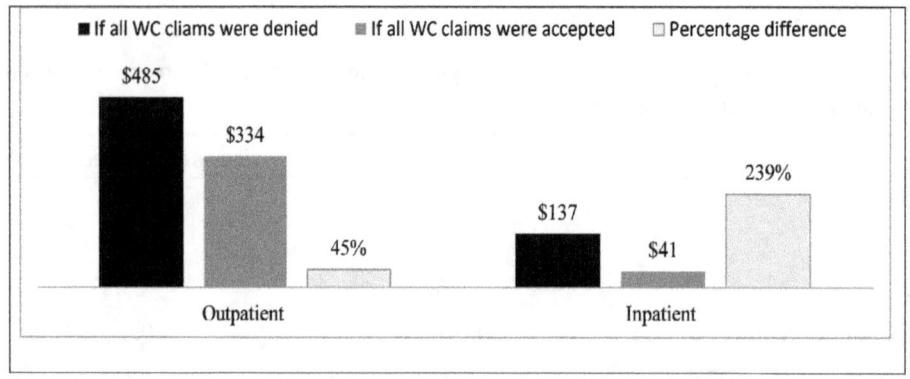

Figure 1. Impact of WC claims denial on inpatient GHI costs within 2 weeks after injury

Occupational Amputations in Illinois: Data Linkage to Target Interventions

Linda Forst, Lee Friedman
University of Illinois at Chicago

Background

Amputations are severe injuries that cluster in certain work sectors and workplaces, disproportionately affect Hispanics and immigrants, and are completely preventable. Amputation injuries are worthy of study because they represent a sentinel injury that is easy to diagnose and rarely disputed as being work-related, making them relatively easy to capture in occupational injury surveillance systems. Furthermore, they horrify everyone, and thus may provide an impetus for enforcers, insurers, policy makers and the public to demand interventions that prevent them.

Illinois, the fifth largest state in the US, has a population of almost 12.9 million and employs some 6 million workers in all economic sectors. Approximately 600,000 people are employed in manufacturing and almost 200,000 in construction, the sectors at highest risk for amputation injury. The goal of this investigation is to determine the numbers, rates, and trends of amputations in Illinois, to compare state-based data with the BLS Survey of Occupational Injuries and Illnesses, to determine the extent of OSHA investigations/citations of known amputations, and to foster a dynamic, statewide intervention program based on surveillance using workers compensation data.

Methods

For this study, we obtained datasets for the years 2000 through 2007 from the Illinois Department of Public Health—the Illinois Trauma Registry (TR); the Illinois Hospital Association via our own hospital—the Hospital Discharge database (HD); and the Illinois Workers Compensation Commission—the "claims" database (WC). We conducted a descriptive analysis of each dataset and linked cases across the three datasets to approximate the total number of cases in Illinois and to describe them. We extracted work-related cases from the TR Registry by designation of either "work-related" or "workers compensation" payer. Five variables were used to do the linkage: age, sex, race/ethnicity (for HD and TR; not captured in the WC database), date of injury, and diagnosis codes. The diagnosis is not coded in the WC dataset, so "amputation" was used as a key word). We compared the number and characteristics of the cases we detected from State databases with the cases that were captured by the BLS-SOII. We reviewed OSHA citations for amputations during the same period to determine how many of the cases found in the Illinois datasets had been investigated. We took these results to State agencies to foster a discussion about potential avenues for prevention.

Results

Table 1 shows the variables collected in each of the three databases. We found a total of 3984 cases in the state databases with only eight cases linking across all three databases; 487 (plus 8) linked across HD and TR; 148 (plus 8) linked across WC and TR; and 10 (plus 8) linked across WC and HD. Amputations by body part were, as follows: thumb 1693 (42.5%), other digits 1522 (38.2%), whole hand 343 (8.6%), forearm 68 (1.7%), shoulder 20 (0.5%), toes 88 (2.2%), foot 64 (1.6%), lower leg 24 (0.6%), and thigh 32 (0.8%). There were data missing about location in 130 of the cases.

Among the 2344 workers compensation claims, 88.8% is male, with 70.4% ranging in age between 25 and 54. Some 54% is married, and almost 37% has dependents--11.2% with three or more dependents. The median weekly wage of these workers was $500 at the time of the injury, with an interquartile wage of $347-$736. While 18% of all workers whose cases go to workers compensation arbitration represent themselves, overall, almost 53% of amputated workers represent themselves.

Where were these workers (from the Workers Compensation database) employed at the time of amputation? The ten employment types with the highest number of amputations, overall, were five temporary employment agencies, the State of Illinois, two food manufacturers, one heavy manufacturer, and one grocery store chain. Employers with the highest number of major amputations were the State of Illinois (8 cases; 5 arm or hand, 3 leg); one temp agency (6 cases, 4 arm or hand, 2 leg); two heavy manufacturers (9 cases; 6 arm or hand; 2 leg, 1 foot); and one waste disposal company (3 cases; 3 arm or hand).

As shown in Table 2, between 2000 and 2007, the BLS SOII estimated that 3637 private-sector, work-related amputations occurred in Illinois. As described above, our analysis identified 3984 cases of amputation during the same period, of which 2.2% were public sector employees. Overall, the two data sources identified nearly the same number of total cases of amputations, with the linked dataset identifying 7.1% more cases. Overall, the amputation cases identified by state-based data sources differed from the SOII estimates by no more than 15%. The biggest differences were seen in 2004 and 2006 with percent differences of 63.1% and 85.8%, respectively.

Illinois is a Federal-plan OSHA state with four OSHA area offices. Between 2000 and 2007, there were 2712 amputation investigations. The top five employers with 20 amputations had a total of 12 inspections. Only three of these five employers were cited for known amputation hazards—lockout/tagout and machine guarding; only one company was investigated and cited within the 60 day statute of limitations. As of 2007, the Illinois Department of Labor provided oversight of governmental employees; any injured workers from federal, state, county, or municipal employers would not be expected to appear in the OSHA citations database. Investigators presented these results to an occupational surveillance advisory board in Illinois, and to the Illinois Department of Labor, which investigates governmental employers and temporary agency employers. The IDOL is looking into the cases to determine how enforcement and other intervention could play a larger role in prevention.

Discussion

Systems designed to capture occupational illnesses and injuries include targeted surveys (eg, BLS SOII), state based workers compensation reports, and rare physician reporting systems. Each of these sources significantly, and often predictably, undercounts the number of work-related illnesses, injuries, and fatalities. There are well known barriers to reporting: on the part of workers, there is underreporting of injuries to supervisors, possibly due to concern about affording lost time or jeopardizing their employment, inability to easily access workers compensation insurance, or an unwillingness to come to the attention of immigration officials; on the part of employers, there may be a disinclination to record incidents in OSHA 300 logs, to report to workers compensation insurers, or to call attention to informal employment arrangements; and there is rare reporting from health care providers who often are unaware that cases are work-related or prefer to access general health insurance rather than workers compensation insurance (Azaroff et al. 2002).

Data linkage allows for capture of the maximum number of cases since it identifies cases present in databases that have different inclusion criteria. Data linkage also makes it possible to fill in missing variables. Finally, data linkage can expand the number and range of variables, thereby offering a more comprehensive picture of demographics, hazards, risk factors, adverse health outcomes, and cost.

The State databases in Illinois were remarkably similar to those of BLS, overall, during the period studied. The significant differences seen in two of the years suggest that the sampling strategy or the weighting of cases should be re-considered. A clearer understanding of how closely the SOII approximates true frequencies on non-fatal occupational injuries and illnesses—which may differ by specific diagnosis—would assist in refining how the sample is handed.

As expected, SOII captures more of the minor injuries, while data from hospitalization and those going to arbitration are likely to be the more severe cases. This speaks to establishing a multi-source system in which the BLS SOII plays a central role. Other possible solutions are to require that all employers report, that the filing is electronic, and/or that the US develop a national survey of workers (Wolfe and Fairchild, 2010).

There is a new proposed rule that requires the reporting of all amputation injuries to OSHA. This change should not only go into effect, but efforts should be made to require reporting from health care providers to OSHA to assure that this public health emergency is addressed. Increased detection of workplace amputations is essential to targeting interventions and to evaluating program effectiveness. This study points out strengths and limitations of the current occupational surveillance systems. It also points out the limitations of at least one workers compensation system to capture cases. Examining current uses of workers compensation data, systematizing data collection, and harmonizing systems across states would add significantly to a national effort to prevent occupational illnesses and injuries across the US.

References
Azaroff LS, Levenstein C, Wegman DH. Occupational injury and illness surveillance, conceptual filters explain underreporting. *AM J Public Health* 2002;92:1421-9

Wolfe D, Fairchild AL. The need for improved surveillance of occupational disease and injury. *J Am Med Assoc* 2010;303(10):981-2

Table 1. Data elements available in Illinois Trauma Registry, Hospital Discharge and Workers Compensation databases.

Database	Inclusion Criteria	Data Elements			
		Demographics	Exposure Data	Health Data	Economic Variables
Trauma Registry (ITR)	Persons treated in level 1 or 2 trauma unit for ≥12 h (~45,000/yr)	Name SSN Gender Age Race/Ethnicity	ICD9 E-codes E849, showing locations where injury occurred Time, day, date of injury	ICD9N & E-codes Body site Severity Hospital procedures Treatment Disability status on discharge Blood alcohol	Cost of hospitalization Hospital procedures Hospital days
Hospital discharge (HD)	All individuals hospitalized in Illinois	Gender Age Race/Ethnicity	ICD-9N and E-codes	ICD9 codes Hospital procedures Hospital cost Discharge status	Cost of hospitalization Hospital days Payer source
Workers Compensation Claims (WC)	Persons filing workers compensation claims for arbitration through IWCC (~70,000/yr)	Name SSN Gender Age	Employer Name Nature of injury Narrative of injury circumstances	Diagnosis key word Hospital procedures Level of disability	Total medical costs Lost wages Cost of compensation Payer source

Table 2. Comparison of amputation injuries of residents in Illinois from state data sources to SOII estimates from 2000 to 2007

Year of Amputation	Total Cases	Private Sector [a]	SOII Estimated Cases	Percent Error [b] Total Cases	Percent Error [b] Private Sector Only
2000	689	674	696	-1.0%	-3.2%
2001	576	563	658	-12.5%	-14.4%
2002	507	496	453	11.9%	9.5%
2003	481	470	540	-10.9%	-12.9%
2004	467	457	280	66.8%	63.1%
2005	471	461	450	4.7%	2.4%
2006	437	427	230	90.0%	85.8%
2007	356	348	330	7.9%	5.5%
Total	3984	3896	3637	9.5%	7.1%

[a] Private sector cases is estimated based on subset of cases with employer information (N=2344), of which 2.2% were employed in the public sector.
[b] Percent error formula: Linked dataset (experimental) minus Survey of Occupational Injuries and Illnesses estimate (accepted value) divided by the accepted value

Copyright 2012 National Council on Compensation Insurance, Inc. Republished with permission. All rights reserved.

The Role of Professional Employer Organizations in Workers Compensation: Evidence of Workplace Safety and Reporting[1]

Harry Shuford
National Council on Compensation Insurance, Inc.

A range of alternative or nontraditional employment arrangements are emerging in US labor markets. One of the most widely discussed in workers compensation is the category of professional employer organizations, often called PEOs. In some sense PEOs involve dual employers—one who controls the worksite and the other, generally the employer of record, who handles a range of human resource services such as payroll, benefits, and workers compensation. The PEO industry often is perceived as a potential, if not actual, problem for the workers compensation industry. Assertions of underreporting claims, misclassifying payroll, and distorting the system of experience rating are common. There is a comparable concern with the quality of the data on workplace injuries collected by the Bureau of Labor Statistics (the Survey of Injuries and Illnesses or SOII.) This short paper provides some analysis of these concerns based on workers compensation data.

The Market Share of PEOs Is Modest

Workers compensation data suggests that the PEO market is highly concentrated. The 15 largest PEO companies comprise approximately two-thirds of the insured PEOs in the workers compensation market in NCCI ratemaking states. Across all 37 NCCI ratemaking states, PEOs account for a relatively small share of all workers covered by workers compensation insurance. In the voluntary market in these states, the PEO share of insured payroll is approximately 1%–2%; in the residual market in the 25 states where NCCI has data, PEOs account for about 4%. PEOs are responsible for a material portion of employment in some states (e.g., FL, TX), especially in the residual market (e.g., AZ). NCCI's data also indicates that on average the worksite employer/clients of PEOs are smaller (under 10 workers) than non-PEO employers (almost 20 workers) (Exhibit 1).

Underwriting Experience of PEO Policies Is Comparable to Non-PEO Policies

Reported frequencies for lost-time claims are typically higher for PEO policies across the three market segments that were examined, including:

1. Voluntary market large deductible policies segment (which typically represent PEO master policies or larger employers)

2. Voluntary market other than large deductible policies segment (which typically are small employers)

3. Residual market segment (primarily small, difficult-to-insure employers).

Exhibits 2 through 4 display results for one of these three market segments[2].

---

[1] Overview of a presentation by Harry Shuford delivered at an NIOSH-organized workshop on "Using Workers Compensation Data for Occupational Safety and Health" June 19–20, 2012, Washington, DC. It was based on the early results of a more extensive study to be published by NCCI in early 2013. Harry Shuford is Practice Leader and Chief Economist at the National Council on Compensation Insurance. His colleague Linda Li was the research project lead, assisted by Eric Anderson.

[2] Results are shown only for the voluntary market and only for insurance policies that are not based on large deductible programs. The specifics vary among the three segments, but the general patterns are comparable. The full set of results will be published once this study is completed.

Copyright 2012 National Council on Compensation Insurance, Inc. Republished with permission. All rights reserved.

Exhibit 1[3]
Comparing PEO Workers Compensation Policies
PEO Client/Employer Appear to Be Small Businesses

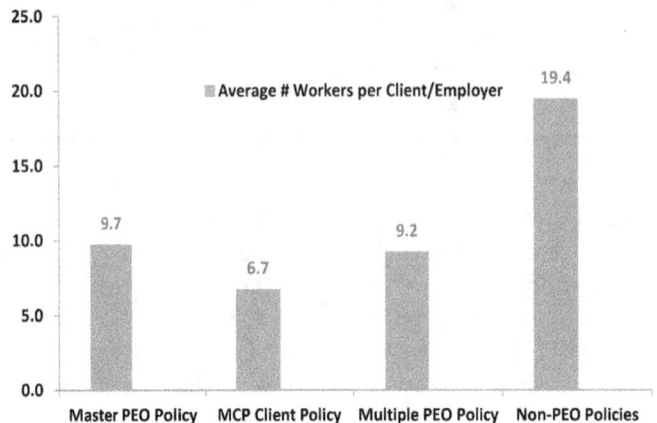

Exhibit 2: Lost-Time Claims Frequency
In the Voluntary Market, Frequency Has Declined But Remains Higher for PEOs Than for Non-PEOs Other Than Large Deductible Policies (Typically Smaller Employers) 2004–2009 at 2nd report

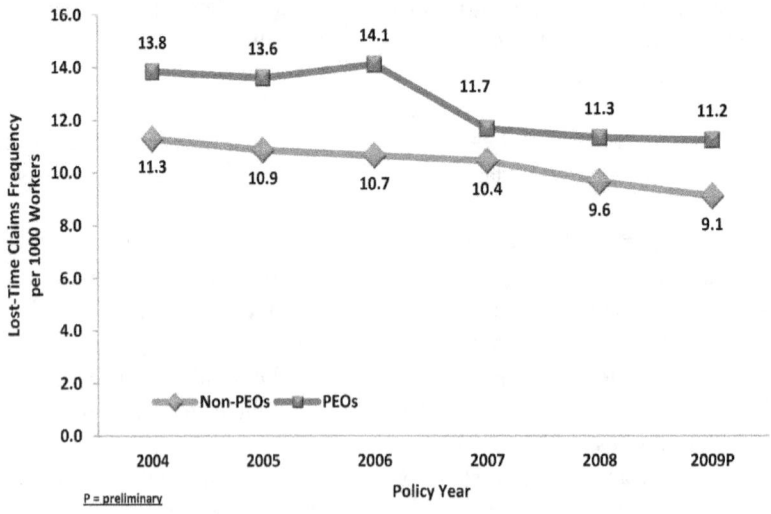

[3] A master PEO policy is a single insurance policy issued to a single PEO but covers "leased" workers for multiple PEO clients. The Multiple Coordinated Policy (MCP) and multiple PEO policies cover the leased workers for a single PEO client.

Exhibit 3: Lost-Time Claims Severity
In the Voluntary Market, Severity Trends Are Comparable But Severity Is Lower for PEOs Than for Non-PEOs Other Than Large Deductible Policies (Typically Smaller Employers) 2004–2009 at 2nd report

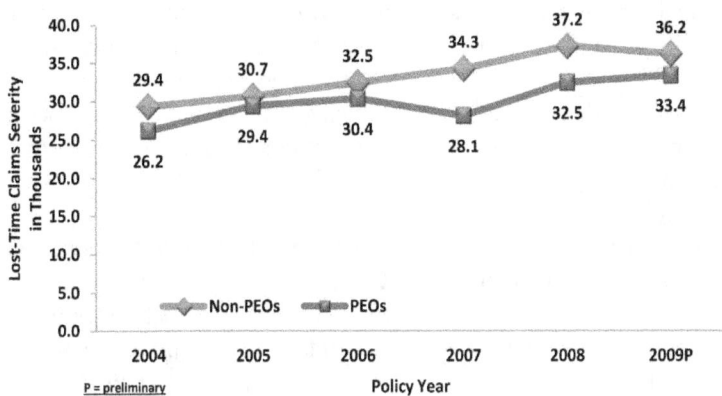

Exhibit 4: Lost-Time Claims Loss Ratios
In the Voluntary Market, Modified Premium Loss Ratio Trends and Levels Are Comparable for PEOs and Non-PEOs Other Than Large Deductible Policies (Typically Smaller Employers)
2004–2009 at 2nd report

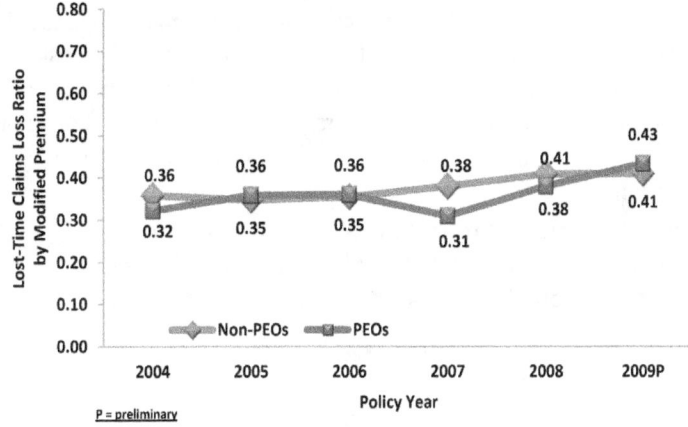

For large deductible policies, PEOs have comparable severity to non-PEOs. PEOs loss ratios based on manual premium have increased over time relative to non-PEOs; however, PEOs modified premium loss ratios (which reflect the application of the experience modification factor) have been lower than non-PEOs. This suggests that on average adverse claims experience is properly reflected in higher experience modification factors. For other than large deductible policies, severity trends are comparable, but levels are lower for PEOs than for non-PEOs, and loss ratios are comparable. In the residual market, severity trends have been higher for non-PEOs than for PEOs, and loss ratios have often been lower for PEOs than for non-PEOs.

The observed higher frequency (Exhibit 2) suggests that underreporting of claims is not a major issue for PEO programs; however, it may indicate the underreporting or misclassification of payroll. The fact that observed severities for PEOs are similar or lower than non-PEO experience (Exhibit 3) is consistent with PEOs reporting claims appropriately. The observation that loss ratios are comparable or lower than for non-PEOs (Exhibit 4) indicates that payroll is typically reported appropriately. Analysis of workers compensation data indicates that there is no material difference in the reporting of workplace injuries and illnesses by PEOs relative to non-PEO employers. The analysis also indicates that on average there is little difference in the overall experience of PEOs and non-PEOs. Moreover, to the limited extent that PEO claims experience is worse than non-PEOs, it is sufficiently embedded in experience modification factors so that the premiums paid by PEOs cover the greater costs.

Mix of Business Across Industry Groups Is Comparable

In the voluntary market, PEOs' industry mix of clients is comparable to the non-PEOs' industry mix (Exhibit 5). PEOs' lost-time claims frequencies are higher for most industry groups than for non-PEOs. PEOs' severities and loss ratios (Exhibit 6) are lower than non-PEOs for all but the miscellaneous group. In the residual market, PEOs are especially prominent in manufacturing compared to non-PEOs (Exhibit 7). PEOs' frequency levels are higher than non-PEOs, while severity and loss ratios (Exhibit 8) are lower in all but the miscellaneous group. Adverse PEOs' experience in the miscellaneous group may reflect the class codes of the employers insured. High risk classes, such as Trucking—Long Distance Hauling—& Drivers, are especially prominent among PEOs in the miscellaneous industry group.

Exhibit 5
In the Voluntary Market, PEO Mix of Clients Is Comparable to Non-PEO Mix Share of Insured Payroll by Industry Group Policy Year 2007 at 2nd Report

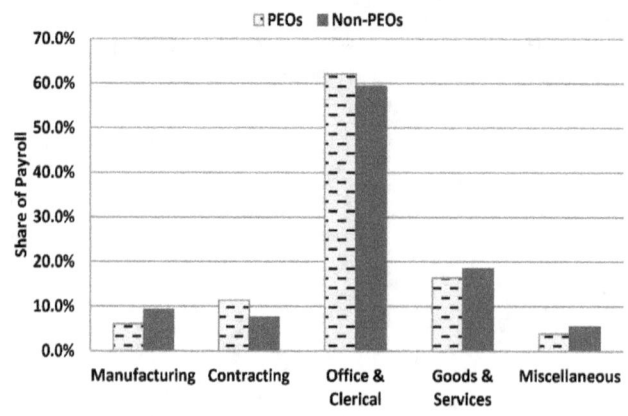

Exhibit 6
In the Voluntary Market, Lost-Time Claims Loss Ratio—Modified Premium PEOs Are Lower Than Non-PEOs for All but Miscellaneous
Policy Year 2007 at 2nd Report

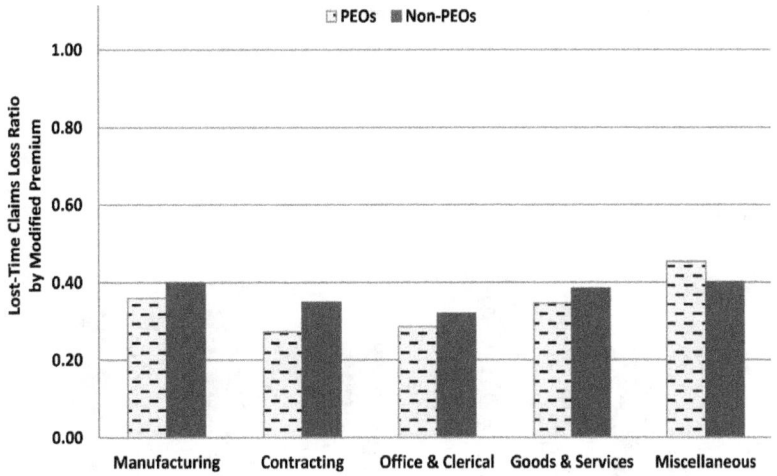

Exhibit 7
In the Residual Market, PEOs Are Especially Prominent in Manufacturing
Share of Payroll
Policy Year 2007 at 2nd Report

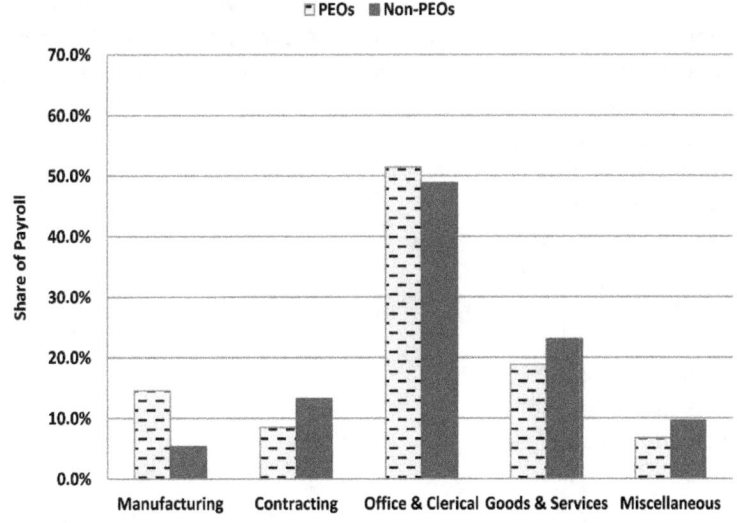

Copyright 2012 National Council on Compensation Insurance, Inc. Republished with permission. All rights reserved.

Exhibit 8
In the Residual Market, Modified Premium Loss Ratios Are Lower for PEOs in All But Miscellaneous
Policy Year 2007 at 2nd Report

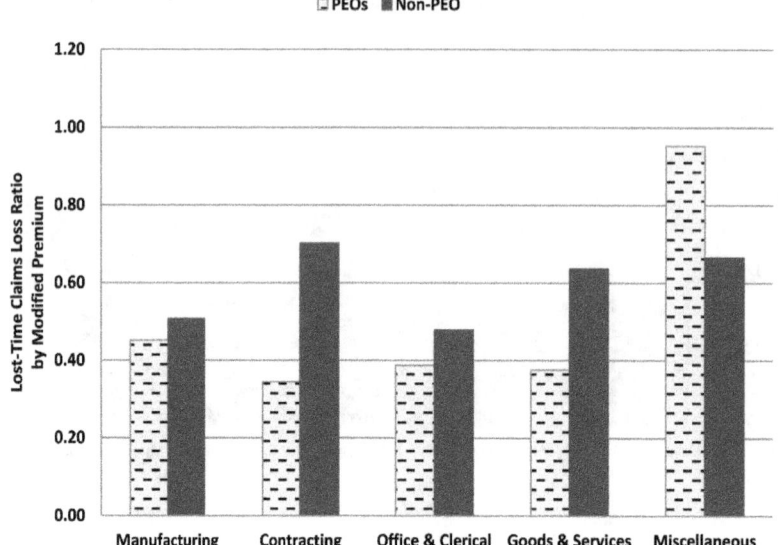

Observations

Differences in workers compensation experience between PEOs and non-PEOS exist but are not as dramatic as many might suspect. Indeed, based on total underwriting experience, as reflected in loss ratios, PEO experience is similar to and in some years superior to the experience of non-PEOs. There are no indications that, in the aggregate, PEO programs exhibit materially worse underwriting experience or that there are material problems with inappropriate reporting. One caveat should be noted: this analysis has not examined the performance of individual PEO programs. As noted in the opening section, the data suggests that as few as 15 national PEOs account for 60% of the insured market; their experience likely dominates the aggregate results reported above. There are several hundred more PEOs that share the remaining 40% of the market.[4] At least a few of these likely have less than stellar results but that is also the case with non-PEO policies.

---

[4] According to the National Association of Professional Employer Organizations (NAPEO), the industry's trade association, there are several hundred PEOs countrywide; this suggests that the majority of them are very small.

Using Workers' Compensation Data to Conduct OHS Surveillance of Temporary Workers in Washington State

Michael Foley, Edmund Rauser, Christina Rappin, David Bonauto
Washington State Department of Labor and Industries, Safety and Health Assessment and Research for Prevention Program

Background

There are several reasons why it may be important to focus OHS surveillance efforts on the portion of the working population without, in the definition of the BLS, "an explicit or implicit contract for long-term employment" [Polivka, 1996]. The share of the total workforce without a permanent employment arrangement is growing; the contingent workforce exhibits several risk factors for injury at higher levels than is true of the permanent workforce; and there are also reasons to believe that injuries to contingent workers are underreported in the BLS Survey of Occupational Injuries and Illnesses (SOII). This project is focused on workers in the temporary help supply (THS) industry as defined within NAICS code 561320. Unlike with other segments of contingent work, (such as direct-hire temps, seasonal workers, and day-laborers) records of hours worked, industries where temporary workers are deployed and counts of workers' compensation claims filed are available in Washington State for workers in the THS industry.

The growth of the THS workforce in Washington State has been rapid as compared to that of the standard-arrangement workforce. Though starting from a small base, the growth rate of the THS industry has averaged 5.0% over the period 1990-2007, compared to the growth in total state employment of 2.3% per year. This growth trend also exhibits a very strong pro-cyclical variation, with a rapid shedding of numbers as the business cycle heads into a recession followed by rapid gains early in the recovery period.

When a worker's tenure at a particular workplace is brief there may be several reasons to expect an increased risk for injury: unfamiliarity with new work practices and surroundings, limited safety training, disproportionately younger workers in this category, or an inability to refuse hazardous work or demand appropriate protective equipment for fear of dismissal. Employers may hire temporary workers as a means of shielding permanent workers from risky tasks, and they may invest less time in providing them with appropriate training and protection equipment. Temporary workers hired through an agency have two separate parties who are responsible for their safety, which raises the possibility that neither will take full responsibility to prepare the worker adequately.

Because temporary help supply workers are the employees of the temporary agency for purposes of payment of wages, benefits and workers' compensation premiums, there may exist in the minds of many client employers the erroneous belief that an injury to a temp worker at the client's worksite should not be recorded on the client employers' OSHA 300 log, which is the source used by employers for completing the survey.

Previous research on the question of whether the rise of temporary or contingent work increases the risk of worker injury has been focused largely on discrepancies in health outcomes rather than on the underlying mechanisms which lead to the differential. Studies have found that temporary workers had higher odds of muscular pain [Benavides et al., 2000]; that in a manufacturing setting temporary workers had injury rates two to three times higher than permanent workers [Morris, 1999]; and that temporary workers had four to seven times the claim frequency compared to permanent workers [Park and Butler, 2001]. In the 2000 European Survey on Working Conditions, temporary agency workers reported greater exposure to physical hazards and a higher level of work intensity and pace than permanent workers [Paoli and Merllie, 2001]. Most studies have not controlled for differences between temporary and permanent workers in their industrial distribution. Data from the 1995 CPS Supplement shows, however, that THS workers are disproportionately concentrated in the manufacturing and services sectors, with relatively low shares in retail and agriculture [BLS, 1995]. Foley [1998], using a large cohort of Washington State workers' compensation claimants, showed that claim frequency and severity as measured by time loss were higher for temporary workers than for permanent workers even after controlling for occupation and industry. Furthermore, this study found that the excess risk increases with the underlying hazard level of the industry. There were similar findings when the analysis was restricted to claims resulting in more than 4 lost workdays. Smith et al [2009] confirmed these results, finding workers' compensation claims rate ratios twofold higher than permanent workers in construction and manufacturing.

Even after controlling for occupation or industry, there remain other sources of difference between temporary and permanent workers which may be associated with increased injury. First among these would be job tenure. Evidence suggests individuals with shorter job tenure are at higher risk for injury or illness [BLS 1997; Breslin, 2006]. Reasons for this association may include unfamiliarity with physical processes and environment, safety procedures and resources [Mayhew and Quinlan, 2002]. Given the much higher percentage of temporary workers who are at the lower range of job tenure, it will be important to separate the independent contribution of job tenure to injury rate from that of employment arrangement. Much the same reasoning applies to the need to control for the age of the worker. The 1995 CPS Supplement found 25% of temporary workers were under the age of 25, as opposed to 15% of permanent workers. As young age has also been associated with injury/illness it will be important to control for this factor as well [BLS, 1997].

In contrast to studies focusing on health outcomes, relatively few studies have examined directly the antecedent factors leading to the discrepant outcomes between temporary and permanent workers. Among these factors may be: To what extent is this difference the result of temporary workers' relative youth as distinct from their brief job tenure? Are temporary workers given the more hazardous jobs in a given worksite? Do they know what to do if they are exposed to hazards? Do they feel unable to refuse unsafe work? What kind of safety training do temporary workers receive at the worksite compared to permanent workers? Do temporary workers underreport injuries more than permanent workers? One study focused on such factors as lack of supervision and training provided to subcontracted employees at a large petrochemical plant which sustained a multiple-fatality explosion in 1989 [Kochan, 1991].

Methods
Workers' compensation data
In Washington State all but the 400 largest employers report hours worked by their employees to the State Fund workers' compensation insurance system. These are grouped by a risk-classification system referred to as the Washington Industrial Classification (WIC) system. The WIC system combines industry and occupation to group workplaces by similar risk of injury for insurance purposes (e.g., a painter and an electrician within the same construction company may have the same NAICS code but will be assigned different risk classes). In all there are 316 "risk classes" in the WIC system, of which 16 are reserved for temporary employees working for temporary help services companies. These include separate classes for office support, technical services, warehousing, retail/wholesale, health care, food processing, agriculture, janitorial services, vehicle operation, machine operators, assembly work and construction. The WIC

system allows us to compare claims rates in each particular kind of temporary work to those of other, permanent-worker risk classes. In order to make valid comparisons, accepted claims received over the most recent five-year period were extracted and the occupation code listed on the claim captured. The distribution of occupational codes (SOC2K codes) on the "candidate" permanent risk class was compared to that of the particular temporary risk class in question. If there was at least a 50% overlap in these codes, the permanent risk class was accepted as a valid comparison to the relevant temporary risk class. For the most recent five-year period we compared accepted claims rates, time-loss claims rates, claim rejection rates, average lost workdays per claim, claim costs and frequency of employer protest between each of the following twelve temporary risk classes and their selected comparable risk classes.

Claimant interviews

Since the overall goal of this surveillance activity is to evaluate the risk factors associated with temporary agency employment we are also conducting telephone interviews with recently injured temporary and permanent workers, matched by workplace and demographic characteristics. These interviews focus on: the nature of the business at which they were injured; the worker's job history; the kinds of tasks they performed; the hazards they faced; how they handle situations they deem to be unsafe; the extent and quality of safety training and equipment provided; the importance of safety to their managers at both the temp agency and the client workplace; the importance of safety to co-workers; suggestions for how to reduce injuries to workers; and suggestions for how best to deliver educational material.

The data collected in these interviews will allow us to test the role played by a number of factors which have been suggested to explain why temporary workers are exposed to higher levels of occupational safety and health risk: their shorter job tenures relative to their permanently-employed counterparts; their relative youth; a lack of sufficient training by either the agency or the client business, as compared to permanent workers; whether they are assigned tasks that are more dangerous than those done by the permanent workers on-site; and whether it is more difficult for them to refuse unsafe work or ask for more training.

We select for interview all workers who filed a time-loss claim in the previous month and whose employer reports hours into one or more of the Temporary Help Services risk classes in the State Fund database. We then select up to three injured standard-employment workers from a comparable risk class as that of the injured temporary worker. We also apply additional matching filters which we believe will make for better comparisons between temporary and standard workers, including age, length of service at the employer and gender.

Results

Workers' compensation claims incidence rates, time-loss days, claim costs per 100 FTE and insurance premium levels were compared to those of permanent employees working in comparable industries and occupations. The selection of comparable permanent risk classes was based upon an analysis of occupational codes listed on workers claims. These comparisons showed that workers who are employed by temporary agencies have a higher claims rate and more lost workdays per 100 FTEs than do their permanently employed counterparts, controlling for industry (Figure 1).

Except in warehousing, health care and office services, claims rates were higher for temporary workers than for permanent workers. The risk ratio for temporary as compared to permanent status ranges from 0.67 to 3.85, with an overall ratio of 1.50. If one excludes the office sector, where manual handling tasks are rare, the overall risk ratio rises to 1.89. Furthermore, there is a positive association between the discrepancy in claims rates and the claims rate for the permanent employed workforce, suggesting that differential exposure to hazards is playing some role in the

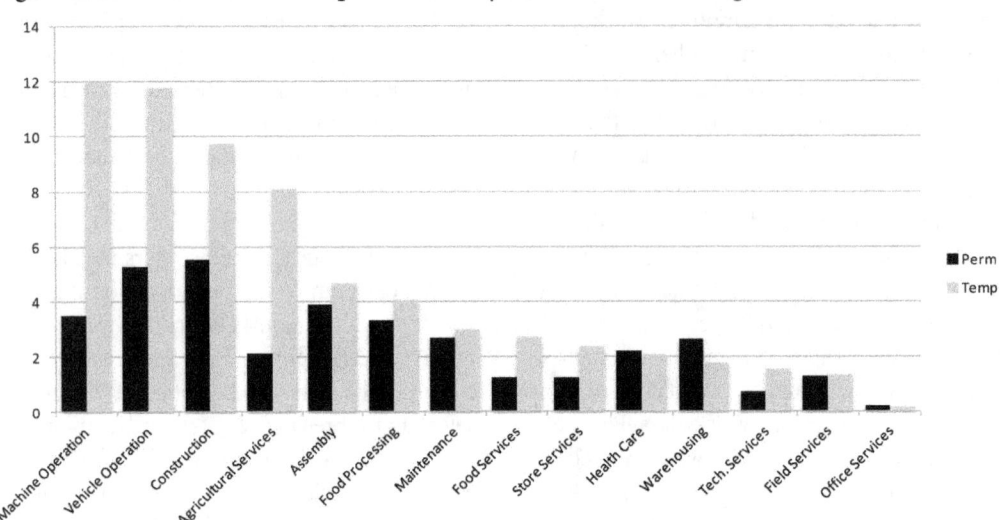

Figure 1: Time-loss claims rates per 100 FTE by Risk Class in Washington State 2005–2011

overall story. A similar discrepancy was found with regard to medical-only claims.

Among the limitations of using workers' compensation data for injury surveillance of temporary workers are that it is well-known that work-related injuries and illnesses are underreported to the workers compensation systems [Fan et al., 2006]. To the extent that there is a differential between temporary workers and permanent workers in claims reporting, estimates of the claims rate discrepancy may be inaccurate. These data are also not able to confirm or refute the role played by differences between temporary and permanent works in specific tasks performed, safety training, or differences in age or tenure. To help with these questions, we conducted telephone interviews with temporary and permanent workers with time-loss claims matched by the procedures outlined above. To date we have completed 34 sets of interviews (each set consisting of one temporary worker and up to three matched permanent workers). The preliminary results show a discrepancy between temporary workers and their permanent counterparts consistent with elevated exposure to risk for the temporary workers. Temporary workers were less likely than permanent workers to report having been asked by their agency or the client about their experience or expertise in the work to be done prior to being assigned: (Figure 2).

They also rated the quality of safety training received by the client employer to be less adequate than that reported by their permanent counterparts: (Figure 3).

Discrepancies suggestive of higher risk for temporary workers were found as well for frequency of training, adequacy of supervision and whether they felt they could refuse tasks they deemed to be unsafe. We found only mixed evidence of discrepancy in level of hazard exposure.

Among the limitations of this kind of study are that it is vulnerable to recall bias, that the very short tenure pattern of temporary workers makes it difficult to match for length of service, and that phone follow-up is more difficult for the younger and more transient workers often engaged in temporary work. It is also expected that we will get a more complete perspective when we have completed the 200 matched sets of interviews envisioned in this project and when we have conducted interviews with managers both at THS agencies and at client businesses.

Figure 2: Asked about experience/expertise prior to job: Temporary vs permanent workers

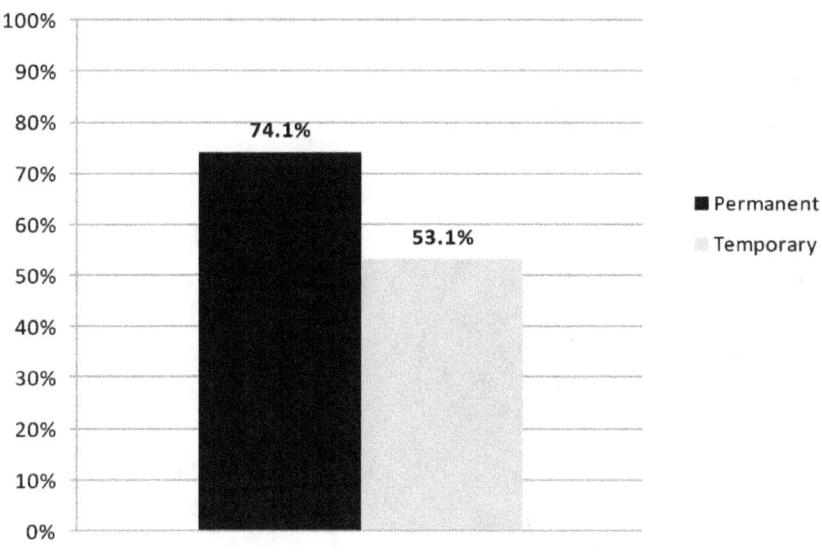

Figure 3: Adequacy of Safety Training

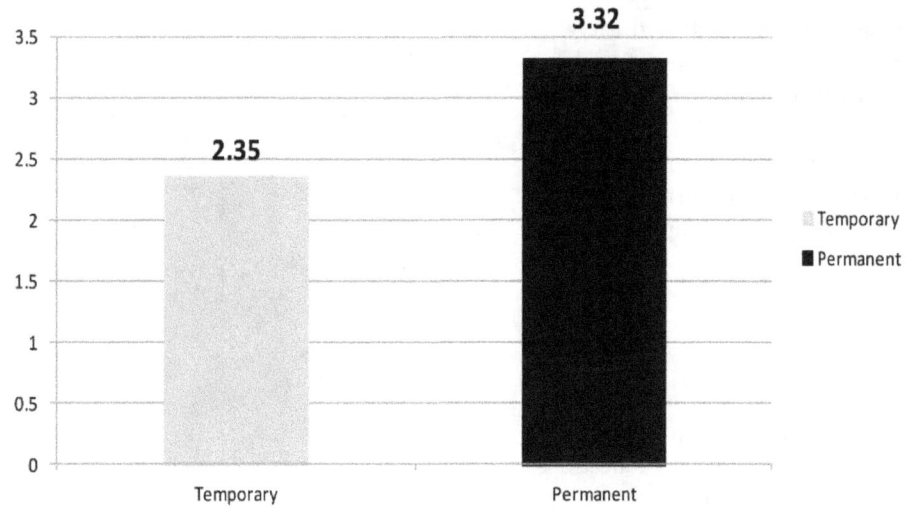

Literature Cited

Benavides FG, Benach J, Diez-Roux AV, Roman C. 2000. How do types of employment relate to health indicators? Findings from the second European survey on working conditions. J Epidemiol Community Health 54:494-501.

Breslin FC, Smith P. 2006. Trial by fire: a multivariate examination of the relation between job tenure and work injuries. *Occup Environ Med* 63:27-32.

Bureau of Labor Statistics. 1996. Current Population Survey, February 1995 Supplement. Washington, DC: United States Department of Labor.

Bureau of Labor Statistics. 1997. Survey of Occupational Injuries and Illnesses, 1995. Washington, DC: United States Department of Labor.

Fan ZJ, Bonauto DK, Foley MP, Silverstein BA. 2006. Underreporting of work-related injury or illness to workers' compensation: individual and industry factors. *J Occup Environ Med.* 48:914-922.

Foley MP. 1998. Flexible work, hazardous work: The impact of temporary work arrangements on occupational safety and health in Washington State, 1991-1996. *Res Hum Cap Dev* 12:123-147.

Kochan TA. 1991. *Managing workplace safety and health: The case of contract labor in the US petrochemical industry.* Dallas, TX: Lamar University Press.

Mayhew C, Quinlan M. 2002. Fordism in the fast food industry: pervasive management control and occupational health and safety risks for young temporary workers. *Sociol Health Illn.* 24:261-84.

Morris JA. 1999. Injury experience of temporary workers in a manufacturing setting: Factors that increase vulnerability. *AAOHN J.* 47:470-478.

Paoli P, Merllie D. 2001. Third European survey on working conditions 2000. European Foundation for the Improvement of Living and Working Conditions. Luxembourg: Office for Official Publications of the European Communities.

Park Y-S, Butler RJ. 2001. The safety costs of contingent work: Evidence from Minnesota. *J Labor Res* 22:831-849.

Polivka AE. 1996. Contingent and alternative work arrangements, defined. *Monthly Labor Review.* 119(10):3-9. Washington, DC: Bureau of Labor Statistics.

Smith CK, Silverstein BA, Bonauto DK, Adams D, Fan ZJ. 2009. Temporary workers in Washington State. *Am J Ind Med.* 53:135-145. doi:10-1002/ajim26728

How WorkSafeBC Uses Workers' Compensation Data for Loss Prevention

Terrance J. Bogyo
WorkSafeBC

Introduction

WorkSafeBC is the exclusive workers' compensation insurer, occupational safety and health regulator, and workplace inspectorate for British Columbia, Canada. It insures 93% of the employed labor force (2.4 million workers) in more than 200,000 firms. Prevention officers in sector-specific industry and labor services, a province-wide field inspectorate of occupational safety and occupational hygiene officers, and a dedicated investigations unit deliver occupational health and safety services. Workers' compensation benefits for work-related injury, illness or occupational disease are payable from the day following the day of injury and include wage indemnity (90% of net earnings), medical costs (WorkSafeBC is the first payer for physician, hospital, medical, physiotherapy, etc.) and vocational rehabilitation.

WorkSafeBC has two other important functions. The first is to act as the regulator of occupational safety and health in the workplace. It's Board of Directors approve the Occupational Health and Safety Regulation, which is then sent for publication by government without the necessity of legislative or executive branch approval.

The other important function relates to premium rate setting. WorkSafeBC uses its own classification system complete with experience rating modification. The rating system is applied to assessable payroll for firms in and about specific industries. The premium rate covers all aspects of workers' compensation including the prevention mandate, health care costs, appeal bodies and advisory services for workers and employers. The Board of Directors announces preliminary mid-year, consults with industry then approves final rates to be effective the following year. This function does not require further approval by the legislative or executive branches of government. The system is funded solely by employer-paid premiums and investment returns on reserves. WorkSafeBC receives no money from governments except for the funds governments pay to cover workers' compensation costs for their employees.

As an integrated, "single solution" to workers' compensation, occupational safety and health, and prevention, WorkSafeBC has workers' compensation data that may be applied to the prevention mandate. Systems are designed with the multiple roles in mind. New systems are increasing the data available for analysis, management and program design in prevention. In the past, performance indicators and statistics were available at an industry sector or province level. For individual firms, claim cost data was also available for premium setting and experience rating purposes. Both of these uses of data were typically available on a quarterly or yearly basis. WorkSafeBC made data accessibility and timeliness of priority for its Business Information and Analysis department. New tools that meet these objectives are now in place. The purpose of this paper is to outline these tools and the approaches adopted to make workers' compensation data a more vital component of the prevention mandate from the corporate management and program design perspective to the enterprise level.

Methods

WorkSafeBC developed and maintains to internal applications that use workers' compensation data for prevention purposes: the Business Planning Toolkit and the Employer

Report Card. Administrators and program management use the Business Planning Toolkit for program design and management. Individual employers and WorkSafeBC prevention staff use the Employer Report Card at the enterprise level to refine prevention efforts and detect patterns of injury that may reveal inherent risks or hazards amenable to improved control through specific interventions. Both applications contain occupational safety and health performance measures (such as injury rate and experience rating). Employers can use these data to assess their individual performance over time and to compare individual employer performance with that of peers in the same classification unit over the same period (thus minimizing the impact common economic and environmental factors that might otherwise influence individual results).

The applications retrieve information from WorkSafeBC's Operational Data Warehouse (ODW). The ODW maintains summarized and detailed copies of information from the different source operational systems (e.g. prevention, claims, assessments) used at WorkSafeBC. The ODW it gets data refreshed daily and monthly and is optimized to allow high performance data analysis through pre-designed self-serve applications.
New data is loaded into the ODW using the traditional "extract, transform, and load" approach which involves the following steps:
- extracting data from outside sources,
- transforming it to fit operational needs, and
- loading it into the end target database.

During the extract phase a series of "change data capture" rules are used to determine the data that has changed in the operation systems, so that it can be loaded into a staging area. This reduces that amount of data that needs to be loaded, ultimately minimizing the amount of time required by the process. The data then undergoes a series of transformation processes in order to meet business and technical needs which may include: translating coded values, deriving new calculated values, joining data from multiple source, aggregating rows, lookup and validate and data. The load phase places the data into the end target database: the ODW.

This data warehouse architecture and processes integrate data from multiple applications, maintain data history, present information consistently and improve data quality over ad hoc extract and linking methods.

The Business Planning Toolkit is an interactive online tool that provides information from the three primary perspectives of WorkSafeBC's mandate: claims, prevention and insurance. It is accessible directly by employers through our secure employer area on the WorkSafeBC website. This application is designed in Microsoft .Net and its data can be accessed 24/7 via an extract from the ODW through web services and it is refreshed monthly.

At the time of writing this article, the Employer Report Card is not currently available online directly to employers but the intent is to include it in the Business Planning Toolkit for release late 2012. At present, WorkSafeBC officers run the Employer report card through an internal reporting portal and provide the report to employers. This application is designed in Microsoft SQL Server Reporting Services and its data can be accessed 24/7 via an extract from the ODW which is refreshed monthly.

Results

The use of the ODW creates opportunities to design "dashboard" applications for routine use, making vital information available on a current basis. Specific data may be portrayed from the perspective of an employer, classification unit, industrial sector or at the provincial level. Key performance measures such as the injury rate, serious injury rate, accepted fatalities, short-term disability claims duration, and fully reserved claims cost (similar to an actuarial incurred cost). The dashboard applications may also be used to cluster or segregate data by characteristics such as employer size, type of injury and age category (e.g. young worker, older workers). Timeframes may be adjusted so that the extracted data reflect trends and cost or injury patterns over time. Presently, WorkSafeBC uses dashboards for two main purposes. The first is to profile claim characteristics (mechanism of injury) by industry, demographics (age/gender), claim type (serious injury), occupation, etc. The second main use of the dashboards is trending performance measures by industry and claim type (e.g., falls). The results are used to identify certain risks for programmed prevention initiatives and specific targeting of prevention and inspection resources.

The application design allows for "drill down" capabilities so that data presented at one view may be examined in detail by clicking on the desired factor. The interface uses icons and indicators common in other applications for ease of use. For example, indicator lights are coded red, yellow and green to represent undesirable, cautionary and desirable states for any specific indicator. A red light indicator at the sector level may prompt a program administrator to expand the sector into its component classification units to identify those pursuits that are contributing to the negative indicator at the sector level. Administrators may use the drilldown feature to identify the firms most responsible for the undesirable performance, important information for targeting prevention resources.

Figure 1 provides a screenshot of the current dashboard view for a specific accident type (Overexertion) over a five-year period at the sector level with an expanded manufacturing sector showing the decomposition of results to specific subsectors. Figure 2 shows a further "drilldown" to the individual firms that contribute to the undesirable outcome at the subsector and sector levels. Note the use of shading (colors in the screen version) to provide a quick visual cue to the specific firms that may be of interest.

Figure 3 takes a different view of data available in the ODW. Here the summary count of short-term disability (STD equivalent to weekly indemnity), long-term disability (LTD equivalent to permanent partial and permanent total disability), and fatality claims (collectively, SLF claims) may be disaggregated to provide details, in this example, by occupation, nature of injury and age category. The interface allows for filtering in any order providing specific groupings that may allow for interventions that are more refined. Figure 4 shows a screenshot with information for a specific employer. The data depict claim cost and experience rating data based on the most recent information. The application allows for intervention scenarios with varying impact to be entered by the user and the impact on experience rating to be projected into the future.

Discussion

Workers' compensation data is a rich source of information for occupational safety and health promotion, targeting and surveillance. WorkSafeBC's integrated mandate to deliver occupational safety and health regulation, workplace inspection, prevention and workers' compensation insurance produces an environment where data from these various functions can be drawn together for these purposes.

The use of an Operational Data Warehouse (ODW) provides a rich data set that combines elements from specific systems into one source that may be accessed by a variety of tools. The ODW approach allows the updating of data on a regular and automated basis increasing the currency of the data for analysis.

To be useful, data must be turned into information. WorkSafeBC has designed specific tools such as the Business Planning Toolkit and the Employer Report Card to access information from the ODW and automate the some complex analysis operations to facilitate data interpretation. By automating the process of extracting and presenting robust, timely information, WorkSafeBC's applications are making workers' compensation data accessible to program administrators, field officers, and individual employers without the need for specialized reports and the resources to create them.

There are limitations to this approach. The data falls short of a real-time dataset because the updating of the ODW is staged at one-month intervals in most cases. The actual impact of the system on decision making at the operational and employer level have not yet been assessed. The utility of data for small employers individually is limited at best because the observational time and number of observations (claims) necessary to detect meaningful trends may have little to do with the current occupational health and safety conditions of the firm today.

Figure 1. Screenshot of current dashboard view [with annotations] for a specific accident type (Overexertion) over a five-year period at the sector level with an expanded manufacturing sector showing the decomposition of results to specific subsectors

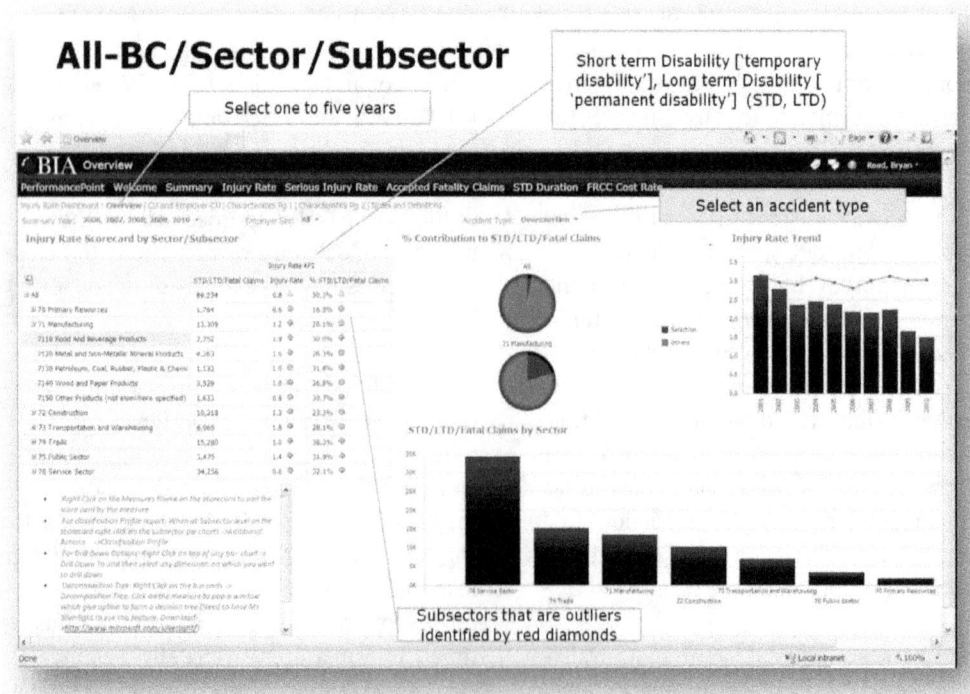

Figure 2. Screenshot of drilldown of dashboard view [with annotations] for a specific accident type (Overexertion) over a five-year period at the sector level with an expanded manufacturing sector and a specific subsector to identify underlying firms

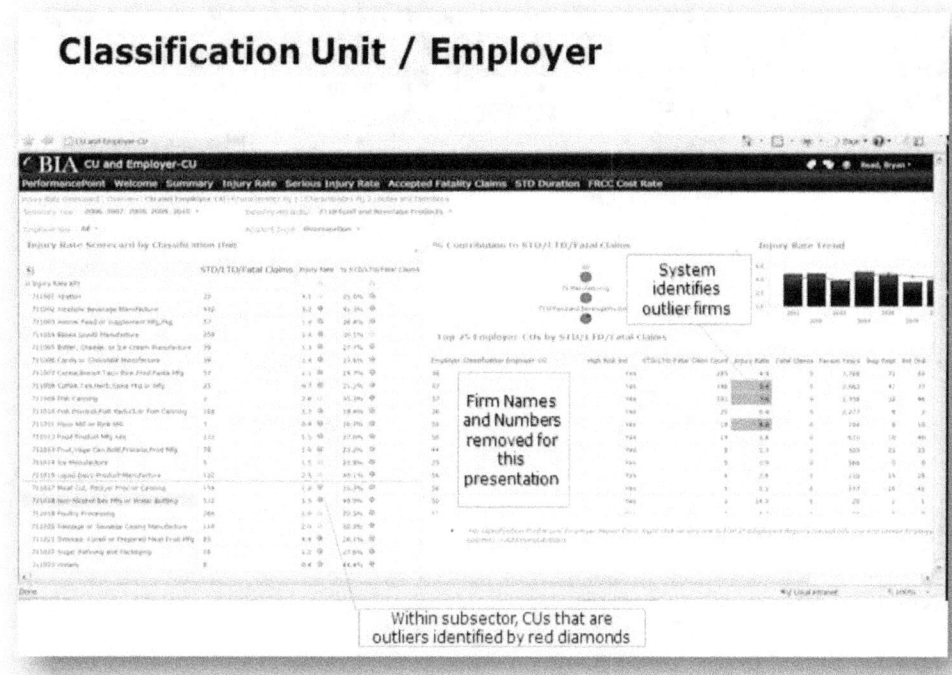

Figure 3. Screenshot showing one branch of the drilldown of short term, long term, and fatal claims (SLF). The order of filters produces clusters and distributions that may be useful in prevention initiatives.

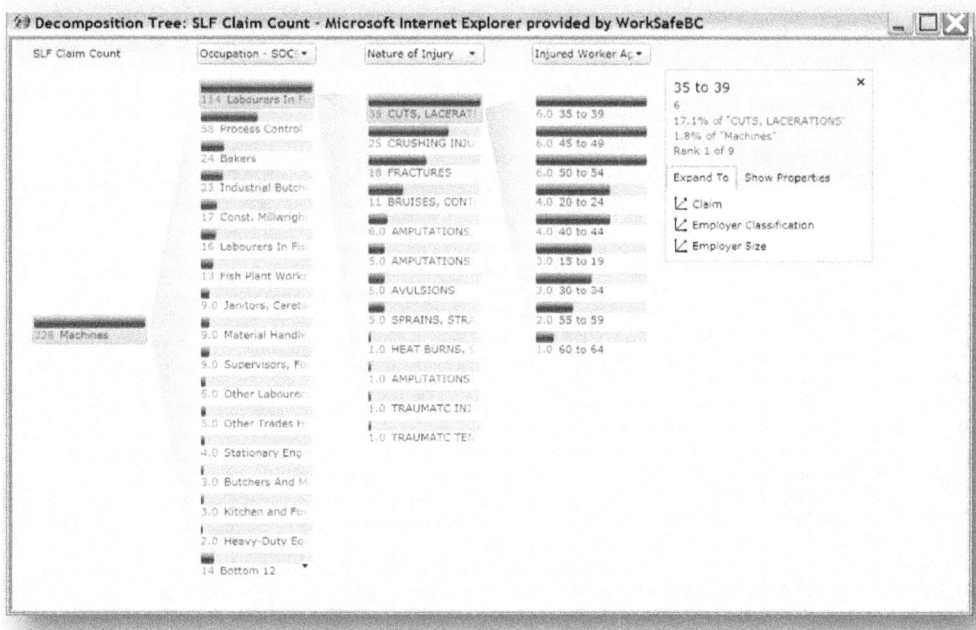

Figure 4. Screenshot showing the experience rating for a specific employer. The filters on the right allow the user to enter a scenario such as efforts to decrease claim costs by 20% and to then view graphically and in financial terms the impact on employer premium costs.

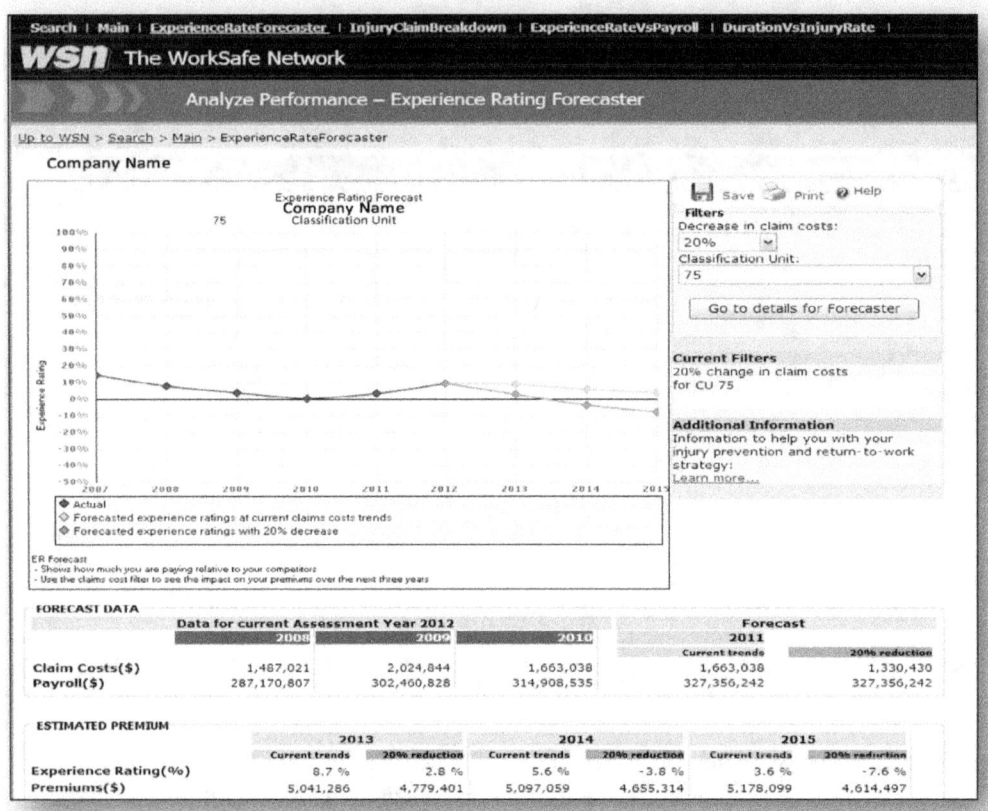

Hitting the Mark: Improving Effectiveness of High Hazard Industry Interventions by Modifying Identification and Targeting Methodology

Christine Baker, Amy Coombe
California Department of Industrial Relations

In an effort to enhance programs that promote compliant business practices, California is actively engaging in strategies that focus on combatting the underground economy. One aspect of this endeavor involves pursuit of egregious violators of labor laws and regulations involving occupational safety and health. In doing so, limited resources are directed in the most effective and efficient manner to produce optimal outcomes. As compliance increases through targeted interventions and general awareness of employer requirements, the environment for workers and businesses in California is enriched and attracts additional economic opportunities.

In the Department of Industrial Relations (DIR), ensuring workplace environments are safe and compliant is a top priority. For purposes of the particular endeavor this paper addresses, the spotlight is centered on identifying businesses with the highest incidence and severity of preventable occupational injuries and illnesses and workers' compensation losses. We consider industry segments where these exposures are acute and regulatory noncompliance is elevated to be "High Hazard". Industry inspections of these high hazard industries result in two levels of citations for violations, including Occupational Safety and Health standards violations and "serious" violations per California Labor Code 6432(a). These levels are correlated with Federal OSHA policy, which assesses state OSH programs based on the percentage of serious citations.

DIR is redefining its strategy for pursuing noncompliance in high hazard industry using a two prong approach: (1) targeting specific industry segments over time to achieve behavioral change, and (2) enhancing the methodology employed for targeted inspections to focus on noncompliant businesses. Using data matching techniques, DIR is targeting specific high hazard industry segments in an effort to curtail noncompliance and incite long-term behavioral change. To monitor progress and evaluate effectiveness throughout the program, key indicators will include violation frequency and percentage of serious violation. If successful, over time we will observe a trend of fewer citations and a decreasing percentage of serious violations in the targeted high hazard industries. These outcomes are indicative of behavioral change and ultimately increased compliance.

Using the concept of targeting high hazard industry segments, our methodology can be captured by a succession of cycles with each subsequent year. By focusing on a specific segment, we anticipate improvement in the indicators described above as expressed over time, until we have exhausted inspections of the empirically-identified "bad actors" in a given segment. Thus the connected cycles that occur with each year of targeted inspections deliver fewer citations and a diminishing portion of serious violations, leading towards greater compliance. For example, if we identified wholesale trade as a targeted segment for this year, we anticipate current inspections would prompt fewer citations next year for not only the businesses that received citations, but also others in the same industry due to awareness and education efforts. We would expect the overall trend in wholesale trade to indicate increased compliance over consecutive years of inspections. Fewer inspections for each segment will be required each year, availing resources to identify and target additional industry segments, such as retail trade, ornamental manufacturing, etc. The cycle would repeat until each segment reached a state of consistent compliance levels.

Sustaining high levels of compliance over time drives behavioral change, which is the ultimate goal of the DIR high hazard program. This methodology is counter-intuitive to standard practices, which support a position of success involving a high frequency of citations and high percentage of serious violations. The refined method suggests that an effective intervention process will produce fewer citations overall when compliance is pursued over time, cultivating an environment conducive to behavioral change.

As illustrated hypothetically in Figure 1, there will be a point of diminishing return when minimal enforcement efforts are necessary to maintain the desired level of compliance within an industry segment. At this point, the yield of citations to inspections is very low and a shift in focus or diversification of resources is merited.

The subsequent step is to identify the next industry segment that will be targeted using the same methodology. It is important to consider resource constraints and external factors that may influence this decision, such as seasonal, economic, and other relevant circumstances.

The second component of the refined DIR targeting methodology involves use of enhanced data matching techniques to empirically isolate noncompliant businesses for targeted inspections. This approach is intended to augment program efficacy by combining the identification process of employers with data on the highest incidence and severity of preventable injuries and illnesses to refine the targeting methodology. We anticipate overall improvement in program outcomes and have planned benchmarks for monitoring and evaluating against baseline data to document the process.

The process we are employing ultimately allows us to develop a more accurate list of businesses to target for inspections. Central to this approach is a new level of data matching where we cross reference high hazard industry segment employers with Workers' Comp Information System data. WCIS supports a universe of data that collects comprehensive information from claims administrators including California workers' comp medical billing data, self-insured and legally uninsured information. It is continuously updated so the information is current, which is invaluable to inspection efforts.

The initial step draws on Workers' Compensation Insurance Rating Bureau (WCIRB) data which has been used historically for high hazard inspection identification purposes. The Days Away, Restrictions and Transfers (DART) data determines annual lists of high hazard industry types, from which the targeted segments are selected. The 2010 California high hazard industry DART threshold is greater than 200 percent, or greater than 4.2. For comparison purposes, the average private sector DART in 2010 was 2.1.

The high hazard industry segment is then cross referenced with Workers' Compensation Information System data. This enables determination of the frequency and severity of the injuries for establishments within the specified industry segment. Through this process we are able to reduce the likelihood of burdening compliant employers with inspections and maximize limited resources by targeting those with the most egregious injury and illness records.

Through this process it becomes possible to prioritize high hazard industry segments based on these criteria and develop a data-driven list of businesses for inspection. The universe depicting this list is the triangle in Figure 2.

Indicators listed below will be monitored to track and assess progress and ensure we move towards the goal of achieving improved workplace health and safety:

- Industry type/segment
- Geography
- Number of violations
- Number of citations
- Percentage of serious/willful violations
- Assessments of violations
- Total number of inspections at various intervals
- Other indicators as needed

This endeavor is significant for DIR because it contributes towards important objectives including: targeting enforcement efforts on bad actors (and not bothering those in compliance); creating a sustainable influence; through inspections; collaborating to increase effectiveness through data matching; developing best practices for enforcement programs; demonstrating program impact through behavioral change; and maximizing resources in austere times.

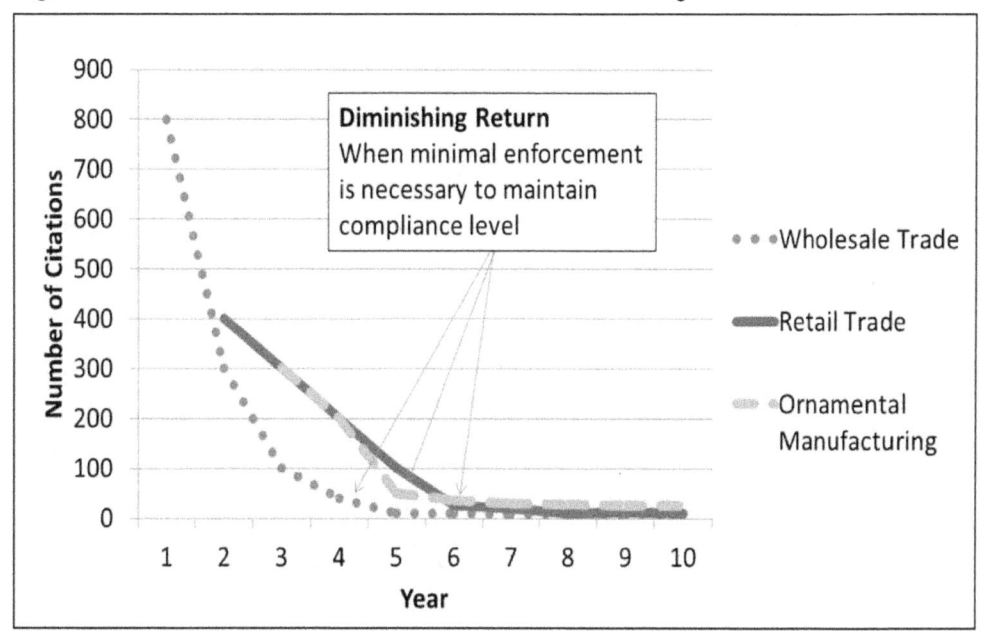

Figure 1. Effective Interventions Will Reach Point of Diminishing Return

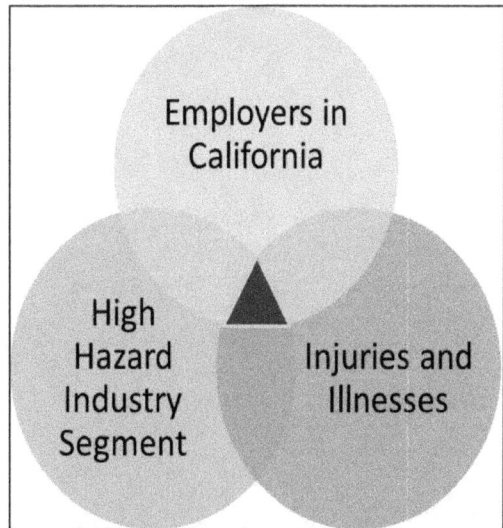

Figure 2. Confluence of Entities Considered to Identify Priority Businesses for Inspection

Injury Trends in the Ohio Workers' Compensation System[1]

Ibraheem Tarawneh, Ph.D., Michael Lampl, MS, CPE, David Robins, Donald Bentley, PE, CIH
Ohio Bureau of Workers' Compensation

Introduction

Ohio is among four remaining states with exclusive workers' compensation (WC) state funds. The majority of Ohio's private and public employers (close to 250,000 employers), employing about 70% of the workforce in Ohio, have their WC coverage through the Ohio Bureau of Workers' Compensation (OBWC). The rest, considered relatively very large employers (close to 2,000 employers) employing about 30% of the workforce are self-insured. OBWC underwrites close to $1.9 billion in premiums annually, which makes it the fifth largest underwriter of WC in the US. The exclusivity characteristic of the Ohio WC system along with the diverse representation of all industrial sectors in Ohio makes the injury and claim data in the OBWC system a great resource for occupational accident and injury surveillance and safety intervention purposes.

While there has been some published research work related to exploring limited portions of the OBWC data (Dunning et. al., 2010; Fujishiro, et. al., 2005 and Marras, et. al., 1995), the data at large, has never been examined and validated for general injury trends. The objectives of this paper are to: 1) Examine and describe general claim of injury trends in the OBWC system over the past decade; and 2) Compare certain claim of injury trends observed in the OBWC system with those observed in similar data sets. The word "injury" will be used interchangeably with "claim of injury" or "claim" throughout the rest of this paper.

Methods

Injury claim data including injury date, ICD-9 codes, injury causation when available, and medical and indemnity costs along with reserves in the OBWC system were gathered for the years 2000 through 2010 to examine emerging trends in frequency and cost of injuries. The data included over 1.8 million injuries that were reported in the OBWC system during those years. Subsets of the data pertaining to lost time injuries for the years 2009 and 2010 were further analyzed to gain better understanding of injury causation and for comparisons with trends reported through other sources, particularly, the Bureau of Labor Statistics (BLS) and Liberty Mutual Workplace Safety Index (2011).

Results

Frequency and Cost of Injuries (2000-2009)

The first observation from the data was that the frequency of reported injuries has steadily decreased for both lost time and medical only injuries over the years 2000 through 2009. Respectively, the number of medical only and lost time injuries went down from 198,337 and 49,427 in 2000 to 82,337 and 20,338 in 2009. When comparing consecutive years, more decreases in the frequency of medical only injuries were observed in the years 2001, 2008 and 2009. On the other hand, more decreases in the frequency of lost time injuries were observed in the years 2006, 2008 and 2009.

The second observation was that although the 30-month medical and indemnity costs for both medical only and lost time injuries

---

[1]The findings and conclusions in this paper are those of the authors and do not necessarily represent the views of the Ohio Bureau of Workers' Compensation

fluctuated from year to year over the ten year period. Those costs peaked in 2002 and 2003 to reach over $648 million and started to go down almost steadily since then.

The third observation was that the considerable decreases in the frequency of injuries did not translate into considerable decreases in the 30-month indemnity and medical costs. For example, a total of 247,764 injuries were reported in 2000 with costs amounting to over $602 million. On the other hand, a total of 129,615 injuries were reported in 2008 with costs amounting to over $546 million. In other words, over nine years, although the number of reported injuries dropped by 47.6%, the 30-month indemnity and medical costs dropped by only 9.3%. These trends are similar to those reported by NCCI: "injury rates had fallen by 56.4% from 1990 through 2009, an average decrease of 4.3% per year" (NCCI Research Brief; August 2011).

Most Costly Injuries

To gain better understanding of these observations, the data was further analyzed according to the "optimal return to work" injury classification diagnosis (ICD-9) codes. The optimal return to work ICD-9 code is defined as: The one allowed ICD-9 code that most likely will keep the injured worker off work for the longest period of disability. For the purposes of the analysis, whenever a claim has more than one ICD-9 code assigned to it, the ICD-9 code that will most likely keep the injured worker off work for the longest time was assigned as an optimal ICD-9 code for that claim. Picking an optimal ICD-9 code among multiple ICD-9 codes assigned to a claim is based on established benchmarks used within the Degree of Disability Measurement (DoDM) to evaluate claim outcomes. When using the optimal ICD-9 code, almost 60% of the injuries reported between 2000 and 2008 fell under a subset of 100 optimal ICD-9 codes. Furthermore, another subset of 100 optimal ICD-9 codes was associated with little over 75% of the total cost of the injuries reported during the same period. Those two trends were revealing as they suggest that optimal ICD-9 codes can be used as a tool for reducing large data sets of WC compensation injuries based on frequency and cost for intervention purposes.

With that premise, the data was further analyzed to examine the type of optimal ICD-9 codes contributing to large portions of the total cost to the system. Part of the analysis is presented in Table 1. Although it was no surprise that the optimal ICD-9 codes described in Table 1 are related to the back, shoulder and knee body parts, the changes of the injury count rank for these injuries over the years was somewhat surprising. The changes to the injury count rank for the four optimal ICD-9 codes are presented in Table 2. Two primary observations about the OBWC system can be made relative to the data presented in Tables 1 & 2 including: 1) Lumbar disc displacement (through their high cost) and sprains to the lumbar region (through their frequency) types of injuries are major drivers of the cost in the system; and 2) Sprain rotator cuff and meniscus of knee current types of injuries are emerging to be major drivers of cost through their frequency and cost to the system. It is worth noting that the majority of the injuries with a lumbar disc displacement optimal ICD-9 code start as injuries with a sprain lumbar region optimal ICD-9 code.

Injury Causation

In July of 2007, the OBWC started coding injury causation for lost time injuries. Accordingly, lumbar disc displacement, sprain rotator cuff, and meniscus of knee current lost time injuries during the last six months of 2007 and 2008 were further analyzed for causation. The results of the analysis are shown in Figure 1. Overexertion and slips/trips/falls (STF) were associated with the causation of 75% to 80% of the injuries with these three optimal ICD-9 codes. It is important to note that these high percentages are partially driven by the fact that lost time injuries in the OBWC system are injuries with eight or more days away from work.

Comparison between Trends in OBWC Data and Trends in the BLS Data and the Liberty Mutual Workplace Safety Index

In an effort to evaluate how the injuries/injury data in the OBWC System compares, in terms of frequency and cost, to the rest of the nation; subsets of the OBWC data were further analyzed and compared with data reported through the BLS and Liberty Mutual's Injury Index. Accordingly, the lost time injuries data for calendar year 2009 were analyzed according to the injury causation. This subset of data included 20,338 injuries. Results from this analysis are shown in Figure 2. Primarily, the results revealed that 30% of the injuries were associated with overexertion and 30% were associated with STFs. On the other hand, days-away injuries caused by overexertion and STFs as reported in the BLS data for 2009 were about 25% each. The numbers in OBWC data are partially driven up by the fact that lost time injuries in the OBWC system are injuries with more than eight days away from work.

The 30 month with reserves cost data for the 2009 lost time injuries was also assembled and analyzed by causation to compare with the 2009 Liberty Mutual Workplace Safety Index. Although there were some differences mostly due to miss-matches between some of the causation categories, the causation based on costs appears to be in synch with those reported in the Liberty Mutual Injury Index. In the OBWC system overexertion, fall on the same level, and fall to a lower level injuries were, respectively, associated with 30%, 16.9% and 15.1% of the total cost. Added altogether those injuries were associated with 62% of the total cost. On the other hand, the Liberty Mutual index reported that overexertion, fall on the same level, and fall to a lower level injuries were associated, respectively, with 25.4%, 15.8%, and 10.7% of the total cost of disabling injuries. The variations between the percentages in the two data sets could be attributed to the fact that the OBWC system does not have a "bodily reaction" causation category, which is part of the Liberty Mutual index. Bodily reaction was associated with 10.5% of the total cost in the Liberty Mutual index. Adding the percentages reported for these four categories in the Liberty Mutual index amounts to 62.4%.

Sub-causation of Overexertion and STF Injuries

In an effort to understand factors leading to overexertion and STF lost time injuries, the 2010 lost time injuries associated with those two causations were further analyzed. This subset of data included 15,367 lost time injuries, of which 32% were associated with STF and 30% were associated with overexertion. Analysis of the overexertion injuries revealed that 38% were associated with lifting tasks and 16% were associated with pushing or pulling tasks. Analysis of the STF injuries revealed that almost 20% were associated with slipping over snow/ice, 12% with slipping over water/grease and 10.4% with tripping over objects.

Conclusions

The exclusivity of the Ohio WC state fund and the diversity of the industrial sectors in Ohio lend the workers' compensation data in Ohio a great value for the purposes of surveillance of occupational injuries and illnesses. However, except for limited research studies dealing with particular types of injuries and workplace safety intervention, the OBWC injury data has not been examined and compared with other occupational injury data sources. The results presented in this paper show the richness of information available through the OBWC data system. Also, the results show that the general trends observed in the OBWC system were consistent with those observed in NCCI, BLS, and the Liberty Mutual Workplace Safety Index. Finally, similar to the rest of the nation, in Ohio injuries caused by overexertion and STFs account for the majority of the cost burden.

References

Dunning, K.; Davis, K.; Cook, C.; Kotowski, S.; Hamrick, C.; Jewell, G.; Lockey, J. 2010. Costs by industry and diagnosis among musculoskeletal claims in state workers compensation system: 1999-2004. *American Journal of Industrial Medicine* 53: 276-284.

Davis, J.; Bar-Chaim, Y. 2011. Workers compensation claim frequency. NCCI Research Brief: August 2011.

Fujishiro, K.; Weaver, J.; Heaney, C.; Hamrick, C.; Marras, W. 2005. The Effect of ergonomic interventions in healthcare facilities on musculoskeletal disorders. *American Journal of Industrial Medicine* 48: 338–347.

Marras, W.; Lavender, S.; Leurgans, S.; Fathallah, F.; Ferguson, S.; Allread, W.; Rajulu, S. 1995. Biomechanical risk factors for occupationally related low back disorders. *Ergonomics* 38(2): 377-410.

2011 Liberty Mutual Workplace Safety Index (2011). Liberty Mutual Research Institute for Safety.

Table 1. Optimal RTW ICD-9 codes associated with considerable total cost of injuries (2000-2008).

Optimal ICD-9 Code	ICD-9 Code Description	Injury Count	Injury Count Rank	Total Cost Rank	Average Total Cost/Injury ($)
722.1	Lumbar disc displacement	16,042	21	1	57,860
847.2	Sprain lumbar region	106,927	2	2	2,761
840.4	Sprain rotator cuff	13,372	26	3	42,656
836	Meniscus of knee current	11,329	29	4	37,672

Table 2. Changes in the count ranking of the Optimal ICD-9 codes associated with considerable total cost of injuries (2000-2008).

Optimal ICD-9 Code	2000	2001	2002	2003	2004	2005	2006	2007	2008
722.1	23	21	20	20	20	20	22	21	21
847.2	2	2	2	2	2	2	2	2	2
840.4	32	28	27	26	24	24	24	23	22
836	40	33	32	28	29	29	28	24	25

Figure 1. Distribution of lumbar displacement, sprain rotator cuff, and meniscus of knee current optimal ICD-9 codes by causation (7/1/2007-12/31/2008).

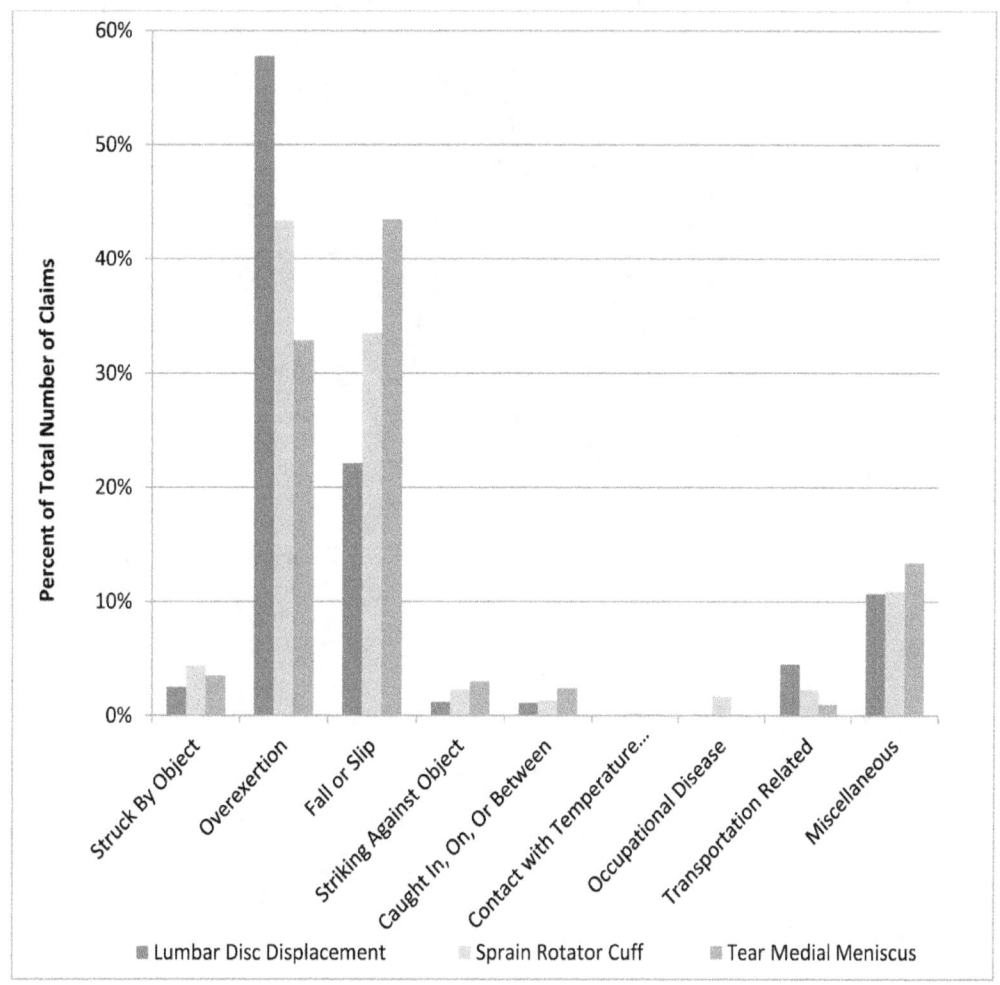

Figure 2. Distribution of OBWC lost time injuries by causation (2009).

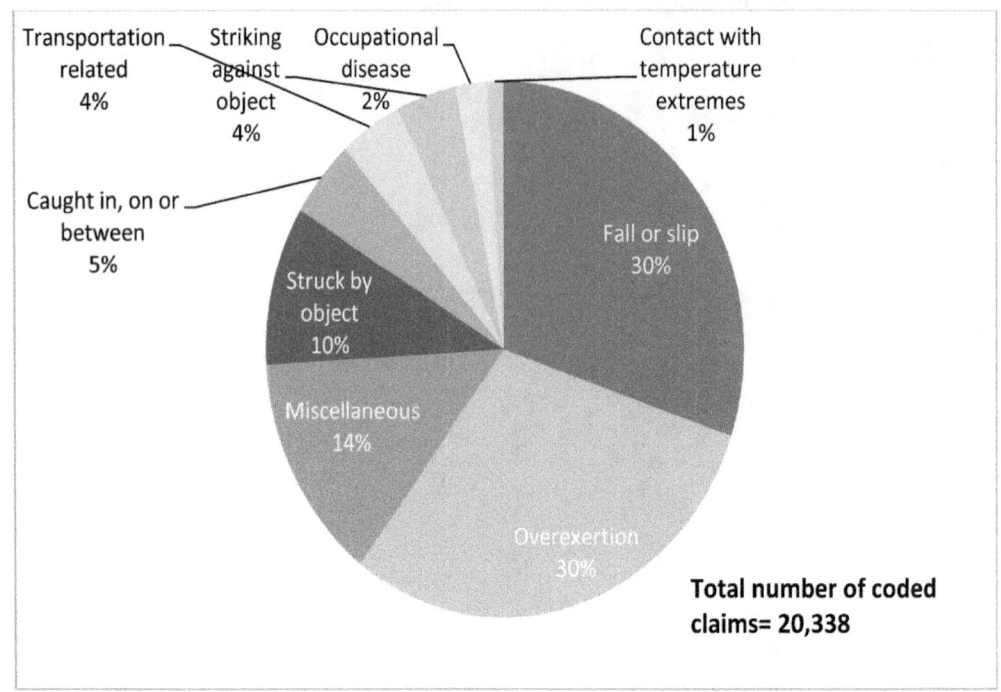

Randomized Government Safety Inspections Reduce Worker Injuries with No Detectable Job Loss

David I. Levine§, Michael W. Toffel*, Matthew S. Johnson#

§University of California, Berkeley, *Harvard University, #Boston University

Abstract

Controversy surrounds occupational health and safety regulators, with some observers claiming that workplace regulations damage firms' competitiveness and destroy jobs and others arguing that they make workplaces safer at little cost to employers and employees. We analyzed a natural field experiment to examine how workplace safety inspections affected injury rates and other outcomes. We compared 409 randomly inspected establishments in California with 409 matched-control establishments that were eligible, but not chosen, for inspection. Compared with controls, randomly inspected employers experienced a 9.4% decline in injury rates (95% confidence interval = −0.177 to −0.021) and a 26% reduction in injury cost (95% confidence interval = −0.513 to −0.083). We find no evidence that these improvements came at the expense of employment, sales, credit ratings, or firm survival.

From: Levine DI, Toffel MW, Johnson MS. 2012. Randomized Government Safety Inspections Reduce Worker Injuries with No Detectable Job Loss. Science 336(6083; May 18): 907-911. Reprinted with permission from AAAS.

Comparison of Data Sources for the Surveillance of Work Injury

Mustard, Cameron A§,*, Chambers, Andrea§, McLeod, Christopher#, Bielecky, Amber§, Smith Peter M§,*

§Institute for Work & Health, Toronto, Canada, *Dalla Lana School of Public Health, University of Toronto, Canada, #University of British Columbia, Vancouver, British Columbia

Introduction

This paper describes the incidence of work injury over a five year period 2004-2008 and has the specific objective to compare the incidence of work-related injury and illness presenting to emergency departments to the incidence of worker's compensation claims in the province of Ontario. In many settings, there are concerns about the reliability of workers' compensation administrative records as a source of surveillance information on the incidence of work-related injury and illness. These controversies center on concerns about the integrity of workplace reporting of work-related injury and illness among particular groups of workers, or for certain types of injuries as well as concerns about some classes of workers (self-employed and independent contractors) who are excluded from insurance coverage (1-7). In describing the concordance between two population sources of surveillance information, the objectives of this study will speak to these concerns.

Both sources of information in this study are population based. In the province of Ontario, citizens are universally insured for medically-necessary health care, including services provided in hospital emergency departments. Similarly, a single publicly-administered insurance agency administers wage replacement benefits and purchases health care services in circumstances of work-related disability. Approximately 30% of the Ontario labour force are in employment relationships that are excluded from coverage by the workers' compensation insurance agency, the Workplace Safety & Insurance Board (WSIB). The WSIB administers work disability claims that result in time off work (lost-time claims) and claims that only require health care services (no lost-time claims). A proportion of both lost-time and no lost-time claimants will seek treatment in a hospital emergency department. In addition, there will be work-related injury or illness episodes presenting to an emergency department that are not reported to, or accepted by, or eligible for coverage from the WSIB.

We evaluated four dimensions of the concordance of the two data sources:
1) the annual rate of change in the frequency of compensation claims and emergency department visits, 2) the ratio of rates of compensation claims and emergency department visits within age and gender groups, 3) the distribution of compensation claims and emergency department visits relative to the external cause of injury, and 4) the incidence of emergency department visits and lost-time compensation claims for serious injuries (defined as those resulting in fracture or concussion).

Methods

The study objective was to compare the incidence of work-related injury and illness presenting to Ontario emergency departments to the incidence of worker's compensation claims filed with the Ontario Workplace Safety and Insurance Board for a complete population of occupationally-active adults aged 15-64 in the province of Ontario over the period 2004-2008. Estimates of annual hours worked for the Ontario labour force by age and gender, derived from labour force surveys, are used to compute rates of work injuries per 2,000,000 hours worked.

Data Sources

Administrative Records of Workers' Compensation Claims

Administrative records maintained by the

Ontario Workplace Safety & Insurance Board contain information describing registered employers and the course and outcome of individual compensation claims. Electronic records of compensation claims resulting in the payment of wage replacement benefits (referred to as lost-time claims in this study) contain information on the date and time of injury, the employer's economic sector and the gender, birth date and occupation of the injured worker. In addition, a national coding standard (CSA Z-795) is used to classify information describing the injury event characteristics and the injury characteristics: 1) the nature of injury; 2) the part of body involved; 3) the source of injury or disease; and 4) the event or exposure (8). Over the period 2004-2008 there were 435,336 lost-time compensation claims.

National Ambulatory Care Reporting System (NACRS)

NACRS was established by the Canadian Institute for Health Information in 1997, providing data on individual client visits to facility-based ambulatory care services, primarily emergency departments in acute care hospitals (9). In July 2000, the province of Ontario mandated the reporting of all emergency department visits to NACRS. There are more than 5 million annual emergency department visits in the province of Ontario recorded in NACRS. For the purposes of this study, we obtained extracts for 707,963 NACRS records reported in the province of Ontario over the period April 2004 to December 2008 with a 'responsibility for payment' code indicating the Workplace Safety & Insurance Board. This coding indicates the clinical determination of a work-related cause of the injury or illness presenting for emergency department treatment and is independent of the registration or acceptance of a workers' compensation claim. Variables included in extracted records were: gender, birth date, visit type, triage date, triage time and a series of up to 10 fields documenting the main problem and the external cause of injury. Of the 707,963 emergency department records, 588,186 (84%) had an accompanying code for an external cause of injury, indicating a traumatic cause.

Measures

Characteristics of the Injury: Two measures were obtained from compensation claim records: 1) nature of injury and 2) part of body injured.

Characteristics of the Injury Event: Two measures were obtained from compensation claim records: 1) source of injury and 2) event leading to injury.

Estimates of annual hours worked: We used information from custom tabulations of the Labour Force Survey to estimate annual hours worked, tabulated in ten year age bands (15-24, 25-34, 35-44, 45-54, 55-64) for men and women separately. Denominator estimates were adjusted for differences in the coverage of the Ontario labour force between the WSIB and the Ontario Health Insurance Plan in the calculation of age and sex specific injury rates (10).

Results

Table 1 reports the distribution of emergency department records attributed to work-related causes and the distribution of accepted lost-time compensation claims for each of the five years in the observation period. Over the five year observation period, the frequency of emergency department visits for work-related causes was approximately 60% greater than the annual incidence of accepted lost-time compensation claims. This ratio was constant over the five year observation period. Between 2004 and 2008, there was a 14.5% reduction in emergency department records attributed to work-related causes and a 17.8% reduction in lost-time compensation claims.

For men, age-specific incidence rates are highest at younger ages for both emergency department visits and lost-time claims (Table 2). In addition, among men the age-specific ratio of the emergency department incidence rate to the workers' compensation incidence rate is highest at younger ages. Among women, the age-specific incidence of emergency department visits declines with age while the incidence of workers' compensation claims rises with age.

For both men and women, the largest proportion of lost-time compensation claims are attributed to injuries arising from 'bodily reaction and exertion (with the single exception of men aged 15-24, for whom the highest proportion of claims are attributed to 'contact with objects or equipment). In contrast, the largest proportion of emergency department visits for both men and women are attributed to injuries arising from contact with objects or equipment (with the single exception of women aged 55-64, for whom the highest proportion of emergency department visits are attributed to falls). For both men and women, falls are responsible for an increasing proportion of both emergency department visits and lost-time claims with increasing worker age.

Approximately 6.5 percent of emergency department visits for men and 5.5 percent for women were attributed to fractures or concussion and for both men and women this proportion rises with age (Table 3). The incidence rates for work-related fracture or concussion injuries were very similar in the two data sources with the incidence rate for men approximately double the incidence rate for women. In both data sources, the incidence rate for fracture or concussion per 2,000,000 hours of work for men is highest at the youngest ages. In contrast, among women, the incidence rate rises with age in both data sources.

Discussion

This study found an important degree of concordance between two potential sources of information for the surveillance of work-related injury and illness. There was strong concordance in temporal trends: between 2004 and 2008, there was a 17.3% reduction in emergency department visits attributed to work-related causes and a 17.8% reduction in lost-time compensation claims. In addition, when restricted to injuries resulting in fracture or concussion, injury incidence per 2,000,000 hours of work by age group and gender was generally similar in the two data sources. When a comparison was restricted to injuries resulting in fracture or concussion, where we would expect urgent care, the study found a strong concordance between incidence rates estimated from emergency department records and incidence rates estimated from workers' compensation lost-time claims.

The study also found some important discordant patterns in the two sources of information on work-related injury and illness. First, young men especially and young women to a lesser degree have a higher incidence of emergency department visits for all conditions than would be expected based on the incidence of workers' compensation claims. Second, a higher proportion of lost-time claims were attributable to non-traumatic musculoskeletal injuries than were observed in emergency department records.

In conclusion, in this setting, emergency department records available for the complete population of Ontario residents are an important source of surveillance information on the incidence of work-related disorders. Occupational health and safety authorities should give priority to incorporating emergency department records in the routine surveillance of the health of workers.

References

1. Boden LI, Ozonoff AL. Capture-recapture estimates of nonfatal workplace injuries and illnesses. *Ann Epidemiol* 2008;18:500-506.

2. Shannon HS, Lowe GS. How many injured workers do not file claims for workers' compensation benefits. *Am J Ind Med.* 2002;42(6):467-73

3. Mustard C, Cole D, Shannon H, et al. Declining trends in work-related morbidity and disability, 1993-1998: a comparison of survey estimates and compensation insurance claims. *AJPH.* 2003;93(8):1283-6

4. Rosenman KD, Kalush A, Reilly MJ, et al. How much work-related injury and illness is missed by the current national injury surveillance system. *J Occup Environ Med* 2006;48:357-365.

5. Seligman PJ, Sieber WK, Pedersen DH, et al. Compliance with OSHA record-keeping requirements. *Am J Public Health.* 1988;78:1218–1219.

6. McCurdy SA, Schenker MB, Samuels SJ. Reporting of occupational injury and illness in the semiconductor manufacturing industry. *Am J Public Health.* 1991;81:85–89.

7. Morse T, Dillon C, Warren N, et al. The economic and social consequences of work-related musculoskeletal disorders: The Connecticut upper-extremity surveillance project (CUSP). *Int J Occup Environ Health.* 1998;4:209–216.

8. Canadian Standards Association. Z-795-96 coding of work injury or disease information. Etobicoke, Ontario, 1996.

9. Canadian Institute for Health Information, National Ambulatory Care Reporting System Manual, 2008-2009 (Ottawa, CIHI, 2008).

10. Smith PM, Mustard CA, Payne JI. Methods for estimating the labour force insured by the Ontario Workplace Safety and Insurance Board: 1990-2000. Chronic Diseases in Canada. 2004;25(3-4):127-37.

Table 1. Comparison of emergency department records for work-related conditions and lost time claims, Workplace Safety & Insurance Board: Ontario 2004-2008

	Emergency department visits for work-related conditions	Lost-time claims, Workplace Safety & Insurance Board	Ratio of emergency department visits to lost-time claims
2004	149,965	94,407	1.59
2005	153,010	93,306	1.64
2006	141,766	86,354	1.64
2007	134,915	83,656	1.61
2008	128,277	77,613	1.65
Total	707,933	435,336	1.61
Percent change: 2004-2008	-14.5	-17.8	

A total of 116 emergency department records were missing information on injury year.

Table 2. Comparison of emergency department records for work-related conditions and lost-time claims by age and gender, Ontario: 2004-2008

			Age			
Males	15-24	25-34	35-44	45-54	55-64	Total
Emergency department records (N)	103,065	123,703	119,513	89,748	36,070	472,099
Annual hours of work (000)	203,197.5	385,844.6	456,725.3	443,150.7	229,300.2	1,718,218.3
Percent of records (row %)	21.8	26.2	25.3	19.0	7.6	100.0
Annual incidence per 2,000,000 hours of work	202.9	128.2	104.7	81.0	62.9	109.9
Lost-time claims (N)	40,240	60,275	74,918	64,641	29,496	269,570
Annual hours of work (000) (1)	159,306.8	288,225.9	355,789.0	355,850.0	180,917.9	1,340,089.6
Percent of records (row %)	14.9	22.4	27.8	24.0	10.9	100.0
Annual incidence per 2,000,000 hours of work	101.0	83.6	84.2	72.7	65.2	80.5
Ratio of emergency department visits to lost-time claims	2.01	1.53	1.24	1.11	0.96	1.37
Women						
Emergency department records (N)	34,337	37,299	43,017	39,248	15,691	169,592
Annual hours of work (000)	169,914.3	297,691.1	339,663.9	341,523.5	150,083.4	1,298,876.3
Percent of records (row %)	20.2	22.0	25.4	23.1	9.3	100.0
Annual incidence per 2,000,000 hours of work	80.8	50.1	50.7	46.0	41.8	52.2
Lost-time claims (N)	18,063	30,902	46,045	48,541	21,431	164,982
Annual hours of work (000)(1)	121,828.6	182,187.0	211,950.3	213,110.7	92,001.1	821,077.6
Percent of records (row %)	10.9	18.7	27.9	29.4	13.0	100.0
Annual incidence per 2,000,000 hours of work	59.3	67.8	86.9	91.1	93.2	80.4
Ratio of emergency department visits to lost-time claims	1.36	0.74	0.58	0.50	0.45	0.65

(1) Annual hours of work are adjusted for age and sex specific worker's compensation coverage estimates

Table 3. Incidence of fracture or concussion, comparison of emergency department records for work-related conditions and lost time claims, Workplace Safety & Insurance Board, Ontario: 2004–2008

	Emergency Department records				Lost-time Claims		
	Fracture or Concussion	All visits	Fracture or concussion as a percent of all visits	Fracture or concussion incidence per 2,000,000 hours of work	Fracture or Concussion	All claims	Fracture or concussion as a percent of all claims	Fracture or concussion incidence per 2,000,000 hours of work (1)
	N	N	%		N	N	%	
Males								
15-24	4,845	103,065	4.7	9.54	4,447	40,240	11.1	11.17
25-34	6,959	123,703	5.6	7.21	6,558	60,275	10.9	9.10
35-44	8,018	119,513	6.7	7.02	8,779	74,918	11.7	9.87
45-54	7,276	89,748	8.1	6.57	6,160	64,641	9.5	6.92
55-64	3,661	36,070	10.1	6.39	3,351	29,496	11.4	7.41
TOTAL	**30,759**	**472,099**	6.5	7.16	**29,295**	**269,570**	10.9	8.74
Females								
15-24	1,098	34,337	3.2	2.58	998	18,063	5.5	3.28
25-34	1,464	37,299	3.9	1.97	1,475	30,902	4.8	3.24
35-44	1,997	43,017	4.6	2.35	2,352	46,045	5.1	4.44
45-54	2,836	39,248	7.2	3.32	3,416	48,541	7.0	6.41
55-64	1,901	15,691	12.1	5.07	2,435	21,431	11.4	10.59
TOTAL	**9,296**	**169,592**	5.5	2.86	**10,676**	**164,982**	6.5	5.20

(1) Annual hours of work are adjusted for age and sex specific worker's compensation coverage estimates

OSHA Recordkeeping Practices and Workers Compensation Claims in Washington; Results from a Survey of Washington BLS Respondents

David Bonauto, Sara Wuellner, Cody Spann, Nicole Reister,
Washington State Department of Labor and Industries, Safety and Health Assessment and Research for Prevention (SHARP) Program

ABSTRACT

Background

Annually the Bureau of Labor Statistics Survey of Occupational Injury and Illness (SOII) publishes state-level occupational injury and illness (OII) estimates. The SOII is currently the only US surveillance system that allows comparisons of OII rates across most US states. SOII relies on employer reporting of OII cases. Employers only report OSHA recordable cases to SOII; thus employer comprehension of OSHA recordkeeping criteria likely influences SOII estimates. Recent research studies using capture-recapture methods estimate the total burden of OII based on the use of state workers compensation (WC) data and SOII data. In order for capture-recapture methods to provide meaningful estimates of the total burden of OII, the data sources must be independent. The degree of source dependence between WC data and SOII is unknown. This presentation provides preliminary data on employer surveys in Washington State regarding OSHA recordkeeping practices.

Methods

We categorized BLS 2008 SOII establishment respondents into groups according to industry, establishment size, geographic location, and preliminary results of a match between Washington SOII DAFW cases and Washington workers compensation data. Within each group SOII establishments were randomly selected for an in-person interview. One hundred thirteen 2008 SOII sampled establishments were ultimately interviewed.

Results

Based on the analyses of 113 respondents, 13 (12%) could not recall or had never completed an OSHA log and likely used workers compensation data for OII recordkeeping. Of the remaining 100 respondents, when asked how they determine if a case is OSHA recordable, 8 (7%) recorded all injuries on the OSHA log, 24 (21%) included all WC claims, 18 (16%) recorded any doctors visit or injury that received medical attention, and 49 (43%) responded that they followed OSHA criteria. Further analysis of the 49 respondents reporting following OSHA recordkeeping criteria, suggest that few demonstrated a meaningful understanding of the criteria for determining if an OII was OSHA recordable and that decision-making involved often involves review of workers' compensation records.

Conclusions

The results from this selected sample of employers suggest poor compliance with OSHA recordkeeping criteria and a significant reliance on information provided through the Washington WC system. If the dependence between OSHA/BLS SOII and state workers compensation data is common across states, the comparison of BLS SOII OII estimates across states is potentially problematic due to the administrative and legal differences in state workers' compensation programs.

Completeness of Workers' Compensation Data in Identifying Work-Related Injuries

Rosenman KD§, Kica J§, Largo T*
§Michigan State University, *Michigan Department of Community Health

Introduction

The administrative data base compiled by state workers' compensation agencies is readily available in many states and has been used to enumerate the annual number of work-related conditions and their costs. However, there is a substantial medical literature that shows that the majority of workers do not apply for workers' compensation; up to half of workers with work-related injuries and an even higher percentage for work-related illnesses (1- 9). The undercount in workers' compensation data has been shown comparing workers' compensation numbers with the numbers identified in medical records (5, 6, 8), surveys of individuals in the general population or with specific conditions (2, 3, 4, 9) and matching names in medical data bases with workers' compensation data (1, 7).

This presentation provides further evidence of the undercount in workers' compensation and compares, the age, gender, race, severity and industry for work related amputations, burns and skull fractures who received wage and/or medical benefits from the workers' compensation system with those who received medical treatment for these conditions but received neither wage nor medical benefits.

Methodology

All 134 acute care hospitals including Veterans' Administration Hospitals in Michigan were required to report work-related amputations, burns and skull fractures. Medical records were reviewed to identify these three work-related conditions treated at a hospital/emergency department (ED) or as an outpatient visit at a hospital based clinic. A case identified using hospital medical records was defined as an individual aged 16 years or older receiving medical treatment at a Michigan hospital/ED/outpatient clinic who had: (a) an amputation-related International Classification of Diseases, Ninth Revision, Clinical Modification (ICD-9CM) diagnosis code: 885.0-.1, 886.0-.1, 887.0-.7, 895.0-.1, 896.0-.3, and 897.0-.7 and the work-related incident occurred at work from 2006-2009; (b) a burn-related ICD-9CM diagnosis code: 940.0-.9, 941.0-.5, 942.0-.5, 943.0-.5, 944.0-.5, 945.0-.5, 946.0-.5, 947.0-.9, 948.0-.9, 949.0-.5; ICD-9CM codes for accidents caused by fire: E890.0-.9, E891.0-.9, E892, E893.0-.9, E894, E895, E896, E897, E898.0-.1, E899) and the work-related incident occurred at work in 2010; (c) a skull fracture ICD-9 diagnosis code (excluding nasal fractures): 800.0-.9, 801.0-.9, 803.0-.9, 804.0-.9. Individuals were contacted via mail and telephone when it could not be determined from the medical record whether the injury was work-related and/or the name of the employer. For amputations, only Michigan residents were included. There were 35 additional work-related amputations treated in Michigan hospitals.

The Workers' Compensation Agency in the Michigan Department of Licensing and Regulatory Affairs provided access to a database of claims for wage replacement due to lost work time. Individuals are eligible for wage replacement in Michigan when they have had at least seven consecutive days away from work (i.e. weekend and five work days). A case identified using Michigan's workers' compensation system was defined as an individual who was in the lost work time wage replacement database with an accepted claim for an amputation (American National Standards Institute (ANSI) nature of injury code 100) that occurred in 2006-2009, a burn (ANSI nature of injury codes 120 or 130) that occurred in 2010, or a skull fracture (ANSI nature of injury code 210 and body part 100, 110,140, 141,146, 148,149, 150, 160, or 198) that occurred in 2010.

Michigan's Poison Control Center (PCC) was used as one source to identify work-related burns, which were defined as an individual for whom a call was made by a burned employee, family member, coworker, or healthcare provider, regarding a consultation of a work-related burn injury in 2010.

The Michigan Fatality Assessment and Control Evaluation (MIFACE) program was used to identify work-related burns and skull fractures that caused death in 2010. This system relies on required reporting by employers to the OSHA State plan's hot line, death certificates and medical examiner reports.

Information from the hospital/ED medical reports, PCC reports and MIFACE reports on each case was abstracted onto a form, including: reporting source(s), payer, type of medical care (hospital, ED, outpatient), hospital name, type of visit, date of admission and discharge, age, gender, race, city and county of residence, employer information (name, address, NAICS code) and injury date. Information unique to the condition was also abstracted, including amputation of multiple digits, first, second or third degree burn, chemical or thermal burn and part of skull fractured. Duplicates identified by more than one reporting source or secondary visits to the same or a different hospital were eliminated, after abstracting all information from every data source where the individual was identified.

Once case ascertainment from medical record review and patient interviews was completed, records in the work-related amputation database were linked to records in the workers' compensation claims database using SAS software, version 9.2 of the SAS System for Windows (copyright 2002-2008 by SAS Institute Inc.). There were several steps in the record-linkage process. First, matches were identified using various combinations of social security number (either all nine digits or the last four digits which often were all that medical records provided), date of injury (or date of hospital admission), first three letters of last name, date of birth, and company name. For cases that matched, the linked record was visually assessed to verify the match. Once this set of matched cases was created, additional matches were sought using less unique information (e.g., patient zip code of residence, date of injury plus/minus thirty days). The matching process was performed on the entire workers' compensation claims database to allow for links to cases not categorized as amputations by that system. For burns and skull fractures all matches were performed manually after merging the data from the non worker compensation sources into the worker compensation file in alphabetical order by last name.

Cases where workers' compensation was identified as the source of payment in a medical record but where the case was not found in the WC wage replacement data base were assumed to have received medical benefits without wage replacement. Finally, WC cases meeting the condition definition that did not match with cases in any of the other data sources (i.e. where WC was the sole source of the case report) were added to the final data base that was used for analysis.

For the analyses comparing the characteristics of individuals who received wage replacement, medical benefits only, or neither (Tables 3 and 4), the grouping for wage replacement only included injuries coded as that injury in the WC data base. This grouping allows comparison to the most common way that WC data is accessed.

Results

From 2006-2009, 2,555 work-related amputations were identified in Michigan. For 2010, 1,885 work-related burns and 114 work-related skull fractures (excluding nasal fractures) were identified in Michigan.

The first column in table 1 shows that 36.1% of the total number of work-related amputations, 16.0% of the total number of work-related burns and 21.1% of the total number of work-related skull fractures identified were coded as these conditions in the Workers' Compensation wage replacement claims data base. After matching the worker compensation data base with the other sources of injury data, 55.4% of the amputations, 17.2% of the burns and 54.4% of the skull fractures were in the wage replacement data base. The additional matches for amputations occurred because an injury that was coded as an amputation in the medical record was coded as a crush, fracture or laceration in the Workers' Compensation data base. For burns, the difference was because of injuries coded as burns in the medical records but coded as multiple injuries or electrical shock in the worker compensation data base. For skull fractures the difference was because of injuries coded as skull fractures in the medical records but coded as multiple injuries or fractures other than skull in the Workers' Compensation data base.

Table 2 shows the percentage of individuals by each of the three work-related conditions with wage replacement, with medical benefits only and who received neither wage replacement nor medical benefits; 22.9% of amputations, 35.2% of burns and 26.3% of skull fractures received neither wage replacement nor medical benefits. The percentage of individuals receiving no workers' compensation benefits who were self-employed was 27.6% for amputations, 5.1% for burns and 23.3% for skull fractures.

Table 3 compares basic demographics and measures of severity for each of the three conditions by workers' compensation status. For amputations, the only significant difference found was that individuals who received wage replacement were the more severe cases, cases involving multiple digits. For burns, significant differences for individuals who received wage replacement were that they were on the average older, had a larger percentage of men and African Americans, had a higher percentage of thermal versus chemical burns, and had a higher percentage of more severe third degree burns. For skull fractures, significant differences for individuals who received wage replacement were that they were on the average younger, had a higher percentage of women and Caucasians, and had a higher percentage with a skull fracture not involving the base or vault of the skull.

Table 4 shows the top five National Occupational Research Agenda (NORA) sectors for each of the three conditions. For amputations, individuals who received wage replacement from workers' compensation were more likely to have been injured in the manufacturing sector and cases from the agricultural sector were absent. For burns, individuals who received wage replacement, manufacturing was the second most important sector as compared to healthcare/social assistance in medical only or no compensation. For skull fractures, individuals who received wage replacement had more injuries in the wholesale retail sector and the healthcare/social assistance sector but agriculture and public safety were absent.

Discussion

The data from Michigan's multi data source surveillance system found that only 55% of work-related amputations, 17% of work-related burns and 54% of work-related skull fractures received wage replacement from the Workers' Compensation system (Table 1). Since many of these injuries had alternative codes in the Workers' Compensation data base, if one had used the worker compensation data base to provide estimates then only 36% of the amputations, 16% of the burns and 21% of the skull fractures would have been identified (Table 1). The large number of injuries missing from the wage replacement data base raises concerns about generalizing findings about these injuries from using the data generally available from worker compensation agencies. Even if access was available to medical only claims, which are not computerized in most states, 23% of the work-related amputations, 35% of the work-related burns and 26% of the work-related skull fractures would be missed (Table 2).

Significant differences between injuries in the workers' compensation system and injuries not in the system included age, gender, race, and severity. Although injuries receiving wage replacement compensation had a higher percentage of severe amputations (amputations of multiple digits), there were a larger number of severe amputations in the medical only and no compensation categories; 109 medical only and 54 neither wage replacement nor medical compared to 98 with wage replacement. Similarly, although a higher percentage of burns receiving wage replacement were more severe (16.3% vs. 3.2% vs. 2.9%), there were a larger number, 23 third degree burns in medical only and 12 third degree burns receiving neither medical or wage replacement versus 21 third degree burns receiving wage replacement (Table 3).

Decisions on which industries to target for intervention would vary depending on which data source was used. Services, healthcare and social assistance and agriculture would receive more attention if non-workers' compensation cases were used for targeting (Table 4).

All the data we used had limitations. We know that the payer in medical records can be inaccurate or change at a later date after more information is obtained by the hospital. For example, we found that in 26% of the charts for amputations, 6% for burns and 30% for skull fractures where workers' compensation was not listed as the payer on the record reviewed that the individual from that medical record could be found in the wage replacement data base. Presumably there are other individuals who received medical only benefits where workers' compensation was not listed as the payer found in the medical chart. We had no way to check on the magnitude of this missing information but presumably it is comparable to the error found with missing information related to wage replacement. We presume that hospital data was missing industry and employment status (i.e. self-employed vs. employed) in a percentage of the records. We were able to address this issue when we matched records with workers' compensation or when we interviewed the injured patients. Finally, the workers' compensation data base was limited to wage replacement with at least 7 consecutive days away from work (i.e. five work days and a weekend) and workers' compensation records were missing race, ethnicity and severity. As with missing information in medical records, we were partially able to obtain some of the missing information when we could match the workers' compensation and medical records and find the missing information in the matching records.

In summary, workers' compensation is not a panacea to address the undercount in the Bureau of Labor Statistics annual survey since it also has a marked undercount. Rather, workers' compensation is a useful component of a multi-source surveillance system that can identify additional injuries and can provide information missing in other sources.

References

1. Biddle J, Roberts K, Rosenman KD, Welch EM. 1998. What percentage of workers with work-related illnesses receive workers' compensation? *J Occup Environ Med*, 40:325-331.

2. Bonauto DK, Fan JZ, Largo TW, Rosenman KD, Green MK, Walters JK, Materna BL, Flattery J; St. Louis T, Yu L, Fang S, Davis LK, Valiante DJ, Cummings KR, Hellsten JJ, Prosperie SL. 2010. Proportion of workers who were work-injured and payment by workers' compensation systems --- 10 states, 2007. *MMWR*, 59:897-900.

3. Katz JN, Lew RA, Bessette L, Punnett L, Fossel AH, Mooney N, Keller RB. 1998. Prevalence and predictors of long-term work disability due to carpal tunnel syndrome. *Am J Ind Med*, 33:543–550.

4. Lakdawalla, D.N., R.T. Reville, and S.A. Seabury. 2007. How does health insurance affect workers' compensation filing economic inquiry, 45(2):286–303.

5. Leigh, J.P., and J.A. Robbins. 2004. Occupational disease and workers' compensation: coverage, costs, and consequences. *The Milbank Quarterly*, 82(4):689–721.

6. Morse T, Dillon C, Kenta-Bibi E, Weber J, Diva U, Warren N, Grey M. 2005. Trends in work-related musculoskeletal disorder reports by year, type, and industrial sector a capture–recapture analysis. *Am J Ind Med*, 48:40–49.

7. Rosenman KD, Gardiner JC, Wang J, Biddle J, Hogan A, Reilly MJ, Roberts K, Welch E. 2000.Why most workers with occupational repetitive trauma do not file for workers' compensation. *J Occup Environ Med*, 42:25-34.

8. Spieler EA and Burton JF, Jr. 2012.The lack of correspondence between work-related disability and receipt of workers' compensation benefits. *Am J Ind Med*, 55(6):487-505.

9. Stanbury M, Joyse P, Kipen H. 1995. Silicosis and workers' compensation in New Jersey. *J Occup Environ Med*, 37(12):1342–1347.

Funding for this project was provided by the National Institute for Occupational Safety and Health Cooperative Agreement U60 OH008466.

Table 1. Percentage Coded/Not Coded as Condition in Workers' Compensation Wage Replacement Data Base in Comparison to Multi Source Data

	In Data Base	
	Coded as Condition	Coded and Not Coded as Condition
Amputations	36.1%	55.4%
Burns	16.0%	17.2%
Skull Fractures	21.1%	54.4%

Table 2. Percentage with Work-Related Condition who Received Wage Replacement, Medical Only or No Workers' Compensation

	Wage Replacement	Medical Only	Neither	Total
Amputations	1417 (55.4%)	554 (21.7%)	584* (22.9%)	2,555
Burns	324 (17.2%)	898 (47.6%)	663* (35.2%)	1,885
Skull Fractures	62 (54.4%)	22 (19.3%)	30*** (26.3%)	114

Self-Employed: *161(27.6%), **34 (5.1%), ***7 (23.3%)

Table 3. Comparison of Work-Related Amputations, Burns, and Skull Fractures by Workers' Compensation Status

	Wage Replacement	Medical Only	Neither	All
Amputations				
Average Age (range)	40.2 (16-75)	38.9 (16-78)	41.0 (16-86)	39.8 (16-86)
≤18 years (%)	1.8	2.9	2.6	2.4
Men (%)	86.9	86.5	90.8	87.6
African American (%)	11.1	11.0	8.1	8.0
Hispanic (%)	7.8	8.1	8.1	8.0
Severity* Multiple Digits (%)	17.0	10.4	9.2	11.8
Burns				
Average Age (range)*	36.6 (16-70)	33.7 (16-78)	33.2 (14-71)	34 (14-71)
≤18 years (%)	2.9	4.9	5.4	4.8
Men (%)*	78.7	58.2	65.3	64.0
African American (%)*	15.5	12.6	12.7	12.9
Hispanic (%)	5.2	4.4	2.2	3.8
Type of Burn (%)*				
Thermal	82.9	70.6	55.2	67.2
Chemical	13.4	23.7	38.3	27.1
Severity (%)*				
1°	6.2	29.0	24.7	25.3
2°	77.5	67.6	72.4	70.2
3°	16.3	3.4	2.9	4.5
Skull Fractures				
Average Age (range)*	39.3 (19-60)	43.5 (20-65)	46.4 (17-75)	44 (17-75)
≤18 years (%)	0	0	2.4	0.9
Men (%)*	70.8	87.8	85.4	83.3
African American (%)*	0	3.7	7.4	5.2
Hispanic (%)	0	3.7	7.4	5.2
Part of Skull (%)*				
Vault	9.5	14.3	7.3	10.5
Base	14.3	79.6	70.7	62.3
Other	76.2	6.1	22.0	27.2

*P<.05

Table 4. Top Five NORA Sectors by Workers' Compensation Status for Work-Related Amputations, Burns and Skull Fractures

Amputations			
Wage Replacement	Medical Only	Neither	All
Manufacture (52%)	Manufacture (49%)	Services (27%)	Manufacture (46%)
Services (20%)	Services (20%)	Manufacture (25%)	Services (21%)
Construction (10%)	Construction (11%)	Construction (18%)	Construction (12%)
Wholesale/Ret (10%)	Wholesale/Ret (11%)	Wholesale/Ret (18%)	Wholesale/Ret (12%)
Trans/Ware/Utilities (3%)	Trans/Ware/Utilities (3%)	Agriculture (11%)	Agriculture (4%)
Burns			
Services (51%)	Services (50%)	Services (55%)	Services (51%)
Manufacture (26%)	Healthcare/Soc Assist (17%)	Healthcare/Soc Assist (14%)	Manufacture (15%)
Wholesale/Ret (9%)	Manufacture (13%)	Wholesale/Retail (12%)	Healthcare/SocAssist (13%)
Construction (6%)	Wholesale/Ret (9%)	Manufacture (12%)	Wholesale/Ret (10%)
Healthcare/Soc Assist (4%)	Public Safety (4%)	Construction (3%)	Construction (4%)
Skull Fractures			
Services (25%)	Services (29%)	Services (27%)	Services (28%)
Wholesale/Ret (21%)	Construction (17%)	Manufacture (21%)	Manufacture (17%)
Healthcare/Soc Assist (17%)	Manufacture (17%)	Construction (12%)	Construction (13%)
Manufacture (13%)	Trans/Ware/Utilities (17%)	Agriculture (9%)	Trans/Ware/Utilities (12%)
Construction (8%)	Public Safety (6%)	Public Safety (9%)	Wholesale/Ret (9%)
Trans/Ware/Utilities (8%)	Wholesale/Ret (6%)	Wholesale/Ret (9%)	

Another Method for Comparing Injury Data from Workers Compensation and Survey Sources[1]

Nicole Nestoriak, Brooks Pierce
Bureau of Labor Statistics

Introduction

The Bureau of Labor Statistics' Survey of Occupational Injuries and Illnesses (SOII) is a widely referenced source of information on workplace injuries[2]. However, there is growing evidence that the SOII undercounts the true number of workplace incidents. Previous research, often based on comparisons of Workers' Compensation (WC) and SOII data, has produced a range of undercount estimates from 30% to 70%[3]. This paper describes aspects of common methods behind this research and offers a complementary approach to measuring the undercount that is less sensitive to state variation in the quality of employer information in WC.

Methods

We define the undercount as occupational injuries and illnesses found in WC data but not in the SOII, once the two data sources are restricted to a common underlying population.

Three Strategies

Besides identifying a common scope, one of the primary difficulties in comparing the SOII and WC is that the SOII is a sample. There are three potential strategies one could use to address the use of the SOII sample to make comparisons between the two data sources. The first option, which we term a "macro" approach, would be to look at case totals from both sources. The SOII data contain sampling weights which allow one to compute a statistically valid estimate of the total number of injuries and illnesses. While this strategy is straightforward, it doesn't account for the possibility that each source is likely to miss some cases making the total undercount greater than the difference in case totals.

A second option, which we term a "micro" approach, would be to match cases and establishments found in the two data sources. One could then restrict the set of WC cases to just those found in sampled SOII establishments. By utilizing the available details to do the case and establishment match, this approach allows a detailed analysis of cases missed by one system or the other. The primary drawback of this approach is the limited information available on firms in WC to link to establishments in the SOII. Any errors in this linkage directly impact the estimated number of cases missed by the SOII.

A third potential approach, suggested below, is a hybrid of the macro and micro approaches. One could use the detailed information to perform a case match but then use all of the WC cases in combination with information on the SOII sample to estimate the number of cases missed by the SOII. This approach avoids the firm linkage, which relies strongly on the detail and quality of data on firms retained in WC systems.

Case Linkage

In both the micro approach followed by Boden and Ozonoff (2008) and the hybrid approach

---

[1] We thank John Ruser for the original idea which motivates this paper and Anthony Barkume, Gwyn Ferguson, Jeffrey Gonzalez and participants at the 2011 National Occupational Injury Research Symposium for comments. This paper relies heavily on Nestoriak and Pierce (2012). Any opinions expressed in this paper are those of the authors and do not constitute policy of the Bureau of Labor Statistics.
[2] For background on the SOII see Selby et.al. (2008) and Ruser (2008).
[3] See for example Boden and Ozonoff (2008), Leigh et al (2004), Rosenman et al. (2006) and Ruser (2008).

outlined below, a first step in measuring the undercount is to link SOII and WC cases using the available detail on worker's name, sex, date of birth, date of injury, the nature of the injury, and some employer information. While a sizable fraction of cases can be linked deterministically, a higher overall match rate can be obtained by employing additional linking strategies. Probabilistic linkage loosens the criteria that all of the fields must agree and places greater weights on fields with unique values. String and numeric comparators make allowances for typos while determining if fields agree. Although determining the quality of case linkage is difficult, it is based on relatively well-defining fields that are similarly represented in the two data sources.

Figure 1 gives a graphical representation of the case linkage. The SOII, since it is a sample, is represented as the smaller circular area while WC cases compose the larger circular area. In figure 1 we imagine that the SOII and WC reflect similar populations but that WC and SOII reporting differences, and SOII sampling, cause the sources to diverge. Based on case linkage, we determine three types of cases: in the SOII but not WC, linked SOII-WC, and in WC but not the SOII. The overlap area reflects linked cases. The upper left area reflects SOII-only cases. These cases may be interesting from a policy perspective (why might injured workers not apply for WC benefits?) but we do not dwell on them here because the micro and hybrid approaches handle them in similar ways.

By way of contrast we do dwell on the final set of cases, those in WC but not the SOII and labeled "WC-only" in figure 1. This final set of cases is not equal to the SOII undercount, because the SOII cases are from a sample. (Cases not sampled by the SOII are still reflected in SOII estimated totals once one applies the sample weights). While some of the WC-only cases are likely at sampled establishments and therefore part of the undercount, many are outside of the SOII sample.

Determining how to divide the WC-only cases into a subset that is in sampled establishments versus a subset that is not differentiates the micro and hybrid methodologies.

Micro method

In the micro approach, the goal is to keep only the WC cases in sampled SOII establishments. Ultimately this requires determining which WC employers were sampled by SOII. One can imagine different protocols for firm linkage that use information from the case linkages, or firm name and address information, or firm identifiers (in particular, federal tax identifiers (EINs) or state unemployment insurance (UI) identifiers), or some combination of this information. We believe firm linkage to be more difficult than case linkage, as key firm linkage fields are not necessarily represented similarly in each data source. Furthermore, for multi-establishment firms the WC case load may reflect the entire firm whereas by design the SOII captures cases associated with sampled establishments. In such situations further adjustments must be made as the WC firm and SOII establishment need not coincide. Identifying these situations, and gauging the lack of coincidence when they occur, is difficult. Finally, state-specific WC systems and reporting requirements may induce nontrivial state variation in the efficacy of the firm linkage.

Firm linkage results are used to keep only the WC cases in sampled SOII establishments. Each remaining WC case is associated with a particular SOII establishment. Figure 2 represents this by excluding from the WC-only set of cases any WC cases not associated with a SOII-sampled establishment via the firm linkage. Furthermore, because each remaining WC-only case has been assigned to particular SOII establishments, any SOII undercount can be shown on an establishment-by-establishment basis. One then applies the SOII sampling weights so as to represent various populations of interest[4]. This highlights one of

---

[4] Researchers often go further and estimate numbers of cases appearing in neither the WC nor the SOII, via capture-recapture analysis (Boden and Ozonoff (2008)); we do not incorporate such estimates here.

the advantages of the micro method: detailed undercount statistics may be calculated by establishment characteristics such as industry or employment size. Such calculations may be valuable for enhancing survey design.

Hybrid method

As mentioned above, the hybrid method begins, just as the micro approach, by linking cases in the SOII to cases in WC using detailed case information. After case linkage, each SOII case is retained with an additional characteristic: has the case also been matched to a WC case? Applying the SOII sampling weights to the linked SOII-WC cases gives an estimate for the population of linked SOII-WC cases.[5] This is, in effect, an estimate for the number of cases common to SOII and WC if the SOII were a census or enumeration, and the case linkage were carried out using the WC and SOII enumeration lists. Subtracting this population estimate of linked cases from the WC totals gives a population estimate for cases missed by the SOII.

Figure 3 represents this method. WC cases are represented as in earlier figures. Rather than showing the SOII cases as unweighted as in earlier figures, figure 3 shows the SOII cases as weighted using sampling weights. For example, the overlapping area in figure 3 shows the population estimate for linked cases (rather than the raw number of linked cases). The non-overlapping areas are population estimates for what each source in turn misses. The area labeled "SOII misses" represents our estimate for the SOII undercount. Because no attempt is made to determine the undercount on an establishment-by-establishment basis, we dispense with the firm linkage of the micro method. While we avoid introducing firm linkage errors in our estimate of the undercount, we are unable to calculate the extent of the undercount by establishment.

Results

We illustrate the hybrid method using two sets of data, from Kentucky and Wisconsin.

The Kentucky WC data are from first reports of injury (FROI) with a date of injury in 2005. In Kentucky FROIs are mandatory for all workplace injuries and illnesses with more than one day away from work. This rule for inclusion is similar to that for the SOII case and demographics data of at least one day away from work beyond the day of injury.[6]

After certain sector exclusions [7], the SOII case and demographics data has 4,333 cases for Kentucky in 2005, which yield a weighted estimate of 24,560 injuries (se=1229). Similar exclusions were made to the WC data and the final case count there was 30,525.

Our case linkage produced 2824 linked cases. Applying sampling weights to these linked, in-sample cases yields a population estimate of 16,030 cases (se=818). Subtracting this estimate from the WC total of 30,525 implies that SOII missed an estimated 14,495 cases. There were 1509 remaining unlinked SOII cases, and applying sampling weights to this set gives a population estimate of 8,530 for the SOII-only portion of cases. Figure 4 updates figure 3 with these numbers and indicates the overall SOII capture rate of 62.9% (se=2.5%). This result would fit in at the low range of the estimates for the states studied in Boden-Ozonoff (2008).

---

[5] This is analogous to, for example, the SOII collecting case-level information on gender of the injured worker and using sampling weights to estimate population totals for injuries by gender. The main difference is that gender is directly collected within the SOII survey instrument rather than inferred via reference to outside data sources.

[6] The FROI data miss some cases (for example, due to failure to report) and also may include some cases that would not be covered by SOII (for example, cases that are not work-related but nevertheless resulted in a FROI filing); these errors will affect our final estimates. However, such errors do not invalidate our general points about linkage methods.

[7] Railroad and Mining sectors are excluded because the data for these industries do not have the detailed person information necessary for matching. The Temporary Help Services industry was excluded as reporting requirements for this industry are different under OSHA and WC rules (OSHA rules govern SOII).

While it is not possible to directly compare results from the hybrid and micro approach using Kentucky data, as we felt a robust firm linkage was not possible, we do make a separate comparison using Wisconsin data for 1998-2001. These results use the data and case linkage from Boden and Ozonoff (2008)[8]. Based on results as reported in their table 2, the micro approach gives a SOII capture rate of 70%[9]. We found the hybrid method gives a modestly higher capture rate of 73.8%. We interpret this to mean that for Wisconsin data any issues with firm linkage did not substantially impact their findings.

Discussion

The hybrid approach described here has advantages and disadvantages relative to the micro approach. The hybrid approach avoids a difficult firm linkage, and may help minimize any impact that cross-state differences in the quality of WC firm data have on estimated SOII capture rates. Furthermore, SOII capture rates may be calculated by characteristic with this method, provided that both data sources contain comparable and relatively error-free common measures (say, month of injury). However, the micro method potentially produces much more information because it allows analysis of the undercount on an establishment-by-establishment basis.

References

Boden LI, Ozonoff A (2008). Capture-recapture estimates of nonfatal workplace injuries and illnesses. *Annals of Epidemiology.* 18(6): 500-506.

Leigh JP, Marcin JP, Miller TR (2004). An estimate of the U.S. Government's undercount of nonfatal occupational injuries. *J Occup Environ Med.* 46(1): 10-18.

Nestoriak N, Pierce B (2012). Comparing injury data from administrative and survey sources: methodological issues. Working paper prepared for the 2012 Joint Statistical Meetings.

Nestoriak N, Pierce B (2009). Comparing workers' compensation claims with establishments' responses to the SOII. *Monthly Labor Review*, May: 57-64.

Oleinick A, Zaidman B (2004). Methodologic issues in the use of workers' compensation databases for the study of work injuries with days away from work. I. Sensitivity of case ascertainment. *American Journal of Industrial Medicine,* 45: 260-274.

Rosenman KD, A Kalush, MJ Reilly, JC Gardiner, M Reeves, Z Luo (2006). How much work-related injury and illness is missed by the current national surveillance system? *Journal of Occupational and Environmental Medicine,* 48(4): 357-365.

Ruser, JW (2008). Examining evidence on whether BLS undercounts workplaces injuries and illnesses. *Monthly Labor Review*, August: 20-32.

Selby PN, Burdette TM, Huband E. (2008). Overview of the Survey of Occupational Injuries and Illnesses sample design and estimation methodology. 2008 Joint Statistical Meetings.

---

[8] Boden, with an agreement from the state of Wisconsin, kindly provided the results of his case match and all of his accompanying programs to the authors.

[9] In later tables they additionally apply adjustments for cases missed by both systems.

Figure 1. Case Linkage

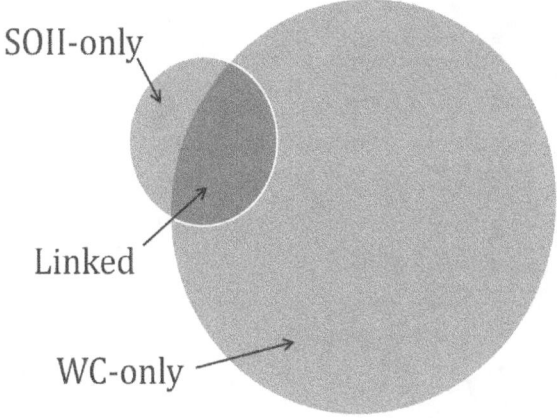

Figure 2. Micro Approach

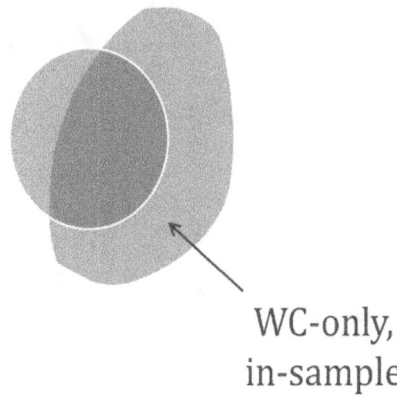

Figure 3. Hybrid Approach

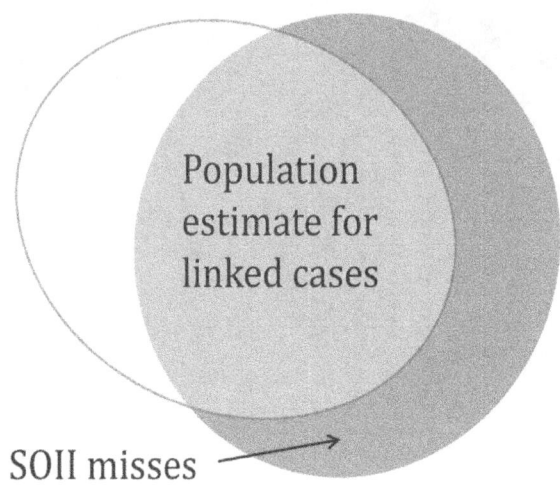

Figure 4. SOII Undercount

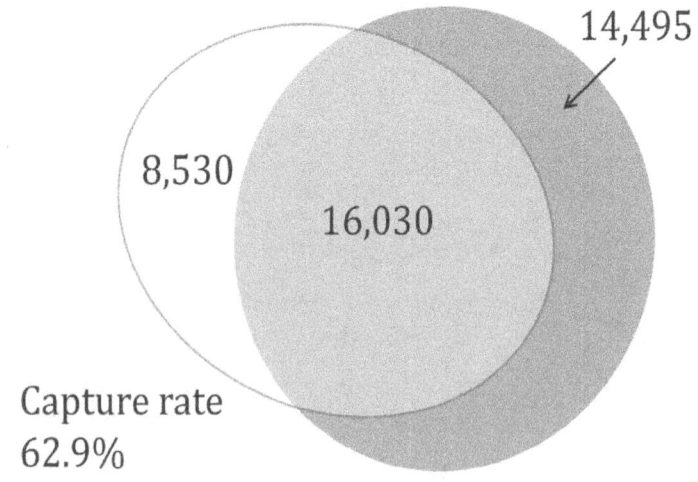

Using O*Net to Study the Relationship between Psychosocial Characteristics of the Job and Workers' Compensation Claims Outcomes

Xiaoxi Yao, MPH; Allard E. Dembe, Sc.D., Cynthia Sieck, Ph.D.
The Ohio State University College of Public Health

Introduction

Extensive research has been performed showing a linkage between occupational psychosocial exposures and health outcomes, particularly with respect to cardiovascular diseases (Schnall et al., 1994) and depression (Netterstrøm et al., 2008). Job psychosocial factors have also been shown to be associated with the duration of work disability following a workplace injury, independently of injury severity and physical workload (Krause et al., 2001). Performing such studies has been difficult because of the dearth of information typically available about workers' exposure to psychosocial risk factors on the job.

For these reasons, some researchers have used the Occupational Information Network (O*Net) data to estimate the risks associated with various job characteristics and worker outcomes (e.g., d'Errico et al., 2007; Cifuentes at al., 2007). O*NET is a publicly available online database that is administered and sponsored by the U.S. Department of Labor/Employment and Training Administration (DOL/ETA). It contains continually updated information on the skill requirements and job characteristics of 974 occupational classifications. Each occupation is characterized by a standardized, measurable set of 277 variables called "descriptors" that describe and rate job requirements, worker activities, workplace conditions, and worker perspectives. The descriptor ratings are based on nearly 70 years of accumulated empirical data collected and analyzed by DOL/ETA. The occupational health research community has become increasingly interested in using O*Net information as indicators of job exposures, especially when other direct measures of workplace exposure are not available.

The primary aim of this exploratory study is to develop and test a generalizable process to use O*Net ratings of job characteristics as surrogate measures of workplace exposures when direct measurement is not possible, and thereby to facilitate studies associating workplace exposures with WC outcomes. We test the utility of that process by assessing seven specific psychosocial job characteristics (e.g., decision latitude, interpersonal relationships.) with respect to their statistical association with three major outcomes in WC claims data (lost work days, claims total indemnity costs and claims total medical costs). We hypothesize that certain job characteristics, such as higher levels of decision latitude, may be associated with better outcomes (fewer lost work days and lower costs).

Only a few related applications of O*Net data for occupational exposure estimation have been conducted previously. Cifuentes et al., (2010) summarized 28 studies in which O*Net estimates of job exposures were used to study health and safety outcomes (e.g., based on survey responses). Only four of the 28 studies based outcomes assessment on WC claims data, and few studies have been done associating O*Net psychosocial variables with WC outcomes. In one study, d'Errico et al. (2007) found decision latitude and supervisor support were significant predictors of the risk of injury. Cifuentes at al. (2007) showed a significant level of agreement between O*Net and questionnaire-based job psychosocial indicators among health care workers.

One of the barriers to such studies is the lack of consistent occupational category coding among WC agencies. Many WC agencies and insurers focus to a much greater extent on claims administration than on accurate attribution of workers' job titles and identifying the inherent risks in particular occupations. To more effectively utilize occupation-specific information in research, this exploratory study aims to develop a generalizable process for using O*Net as a job exposure matrix that can associate occupational exposures with various workers' compensation health and economic outcomes. The proposed analytical process has broad relevancy and wide potential applicability across diverse occupations, job activities, workplace characteristics, and WC outcomes.

Methods

We used a de-identified sample of 316,916 closed Ohio WC claims from 2008 and 2009. We identified 41 specific occupational category titles from WC claims, consisting of all occupation categories containing at least 1,000 claims. This totaled 182,390 claims, which accounted for 57.6% of all claims. O*Net's "Occucoder" software (see: http://www.itsc.org/pages/pub_aocode.aspx) was used to select the closest O*Net occupational codes to map onto each of the 41 WC occupational categories identified by WC claim coding. We then obtained the O*Net ratings for each of seven psychosocial job descriptors in each of the 41 O*Net codes. O*Net assigns an intensity level (on a scale of 0 to 100) for each of those psychosocial characteristics in every occupation code.

> The seven O*Net psychosocial job descriptors we selected for this study were:
>
> *Freedom to Make Decisions* -- How much decision making freedom, without supervision, does the job offer?
>
> *Time Pressure* -- How often does this job require the worker to meet strict deadlines?
>
> *Work schedules* -- How regular are the work schedules for this job?
>
> *Establishing and Maintaining Interpersonal Relationships* -- Developing constructive and cooperative working relationships with others, and maintaining them over time.
>
> *Communicating with Supervisors, Peers, or Subordinates* -- Providing information to supervisors, co-workers, and subordinates by telephone, in written form, e-mail, or in person.
>
> *Scheduling Work and Activities* -- Scheduling events, programs, and activities, as well as the work of others.
>
> *Time Management* -- Managing one's own time and the time of others.

Workers were categorized into high and low exposure groups according to the O*Net intensity level ratings for every job characteristic. For example, assume that a worker's occupational classification is a "human resources manager," and the O*Net level rating regarding "freedom to make decisions" for human resources managers is 91 out of 100 in that occupation. Assume further that the average (mean) level rating for that characteristic (i.e., freedom to make decisions) is 73. In that case, the worker would be placed into the high exposure group for "freedom to make decisions" because the level in his job (91) is greater than the mean rating (73) for that characteristic (freedom to make decisions) across all job classifications.

Once the workers' were classified into high and low exposure groups for each of the seven psychosocial job characteristics, as described above, then the analysis was performed. The independent variable was a bivariate classification of high or low intensity level in each psychosocial job characteristic. The dependent variables in the analysis were: lost work days, claims total indemnity costs and claims total medical costs. Each of those variables was also bifurcated into high and low levels for each outcome. We used the mean value to distinguish between "high" and "low" outcome values. In the case of lost work days, the data was highly skewed (that is, a small proportion of claims accounted for a large proportion of the lost work days). In that case, neither the mean nor the median of the data was considered to be an appropriate cutoff value. To achieve relatively comparable numbers in each group, we thus designated the cutoff as being 7 or fewer days for inclusion in the low group, and 8 or more days for inclusion in the high group. Each logistic regression analysis was then performed separately using a combination of a particular job characteristic intensity level and a particular outcome variable for each worker.

Results

As indicated in Table 1, every psychosocial job characteristic was significantly associated with lower total indemnity costs, except for time pressure, which was associated with higher costs. Time pressure, compared to the other factors, is generally considered to be an undesirable job feature, which means that each of the findings in the table below consistently suggested a relationship in the same direction (e.g., less time pressure is associated with lower costs, consistent with the other findings).

Five of seven findings in Table 2 are statistically significant. But all the calculated odds ratios appear to be clustered closely around 1.0, indicated little true effect, and thus little influence on total medical costs.

As indicated in Table 3, all seven of the psychosocial characteristics are significantly and positively related to the mean number of lost work days. Although higher O*Net psychosocial ratings consistently appear to be associated with more lost days, the reasons for this trend are not clear entirely. It may be that people employed in better psychosocial work environments are more likely to report injuries and work disability than are workers in inferior work environments (i.e., inferior with respect to the level of psychosocial job characteristics).

Discussion

Workers' compensation research is currently impeded by the lack of reliable information about workplace attributes and job functions across occupational classifications. This lack of information makes it difficult for organizations (e.g., employers, WC insurers and state WC agencies) to identify hazards in specific occupations and develop effective strategies for optimizing worker outcomes. Data on workplace exposures in particular occupations is often absent or difficult to obtain. This is particularly true for psychosocial exposures,

Table 1. Occupational Psychosocial Characteristics and Total Indemnity Costs

	Total Indemnity Costs		
	OR	CI	P-value
Freedom to make decisions	0.67	(0.64, 0.71)	<0.001
Time pressure	1.23	(1.16, 1.31)	<0.001
Work schedules	0.90	(0.85, 0.95)	<0.001
Establishing and maintaining interpersonal relationships	0.73	(0.69, 0.77)	<0.001
Communicating with supervisors, peers, or subordinates	0.69	(0.66, 0.73)	<0.001
Scheduling work and activities	0.73	(0.68, 0.77)	<0.001
Time management	0.68	(0.64, 0.73)	<0.001

Table 2. Occupational Psychosocial Characteristics and Total Medical Costs

	Total Medical Costs		
	OR	CI	P-value
Freedom to make decisions	0.98	(0.96, 1.01)	0.163
Time pressure	1.07	(1.04, 1.10)	<0.001
Work schedules	1.07	(1.04, 1.10)	<0.001
Establishing and maintaining interpersonal relationships	0.91	(0.89, 0.93)	<0.001
Communicating with supervisors, peers, or subordinates	0.93	(0.91, 0.96)	<0.001
Scheduling work and activities	0.94	(0.92, 0.97)	<0.001
Time management	0.97	(0.95, 1.00)	0.038

Table 3. Occupational Psychosocial Characteristics and Lost Work Days

	Lost Work Days		
	OR	CI	P-value
Freedom to make decisions	1.25	(1.23, 1.28)	<0.001
Time pressure	1.21	(1.18, 1.24)	<0.001
Work schedules	1.24	(1.21, 1.27)	<0.001
Establishing and maintaining interpersonal relationships	1.30	(1.27, 1.33)	<0.001
Communicating with supervisors, peers, or subordinates	1.31	(1.28, 1.34)	<0.001
Scheduling work and activities	1.24	(1.21, 1.27)	<0.001
Time management	1.19	(1.16, 1.21)	<0.001

for which systematic assessment of job risks is rarely conducted. This study is one of the first studies to specifically analyze the relationship between specific psychosocial factors and WC outcomes using O*Net hazard indicators.

More generally, this study illustrates the potential benefits of utilizing O*Net ratings of job characteristics in various occupations to help facilitate research about workers' compensation outcomes. In some circumstances, it may be advantageous to derive surrogate psychosocial exposure estimates by relying on the ratings of job characteristics contained within the O*Net system.

Deriving hazard estimates by using O*Net occupation-specific ratings can potentially help fill the gap created by the inability to collect exposure information in traditional ways. It potentially allows WC officials and other stakeholders to anticipate the key risks facing workers in various occupations and to plan preventive measures accordingly. Moreover, the ability for occupational job titles available within O*Net to be mapped onto other existing coding systems -- e.g., state WC databases and the Standard Occupational Classification (SOC) coding system - further enhances O*Net's utility.

A limitation of this technique is that the O*Net levels are only indirect estimates of actual exposure. Nonetheless, the many decades of DOL/ETA's empirical data collection that underlies the use of O*Net data as surrogates for actual exposure measurement, provides a wealth of accrued knowledge about various job characteristics that can support this type of analytical process.

Preliminary results from this study suggest that there are statistically significant associations between higher psychosocial factor O*Net ratings and lower indemnity costs. Similar effects were not observed for medical costs. However, ironically, higher O*Net psychosocial ratings appear to be consistently associated with more lost days. The reasons for this trend are not entirely clear. We did not have data pertaining to wage rates available in this study. But it could be that individuals with jobs having positive (desirable) psychosocial attributes work in higher wage positions with less "rush" or pressure to return to the job quickly following an injury or illness. Additional research is needed to better understand the complex relationships that exist between these psychosocial work characteristics and WC outcomes.

References

Cifuentes M, Boyer J, Gore R, d'Errico A, Tessler J, Scollin P, Lerner D, Kriebel D, Punnett L, Slatin C. 2007. Inter-method agreement between O*NET and survey measures of psychosocial exposure among healthcare industry employees. Am J Ind Med 50:545–553.

Cifuentes M, Boyer J, Lombardi DA, Punnett L. 2010. Use of O*NET as a job exposure matrix: A literature review. *American Journal of Industrial Medicine* 53(9): 898–914.

d'Errico A, Punnett L, Cifuentes M, Boyer J, Tessler J, Gore R, Scollin P, Slatin C. 2007. Hospital injury rates in relation to socioeconomic status and working conditions. *Occup Environ Med* 64:325–333.

Krause N, Dasinger LK, Deegan LJ, Rudolph L, Brand RJ. 2001.Psychosocial job factors and return-to-work after compensated low back injury: A disability phase-specific analysis. *American Journal of Industrial Medicine* 40(4): 374–392.

Netterstrøm B, Conrad N, Bech, P, Fink P, Olsen O, Rugulies R, Stansfeld S. 2008. The Relation between Work-related Psychosocial Factors and the Development of Depression. Epidemiol Rev 30:118–132.

Schnall PL, Landsbergis PA, Baker D. 1994. Job Strain and Cardiovascular Disease. *Annual Review of Public Health* 15: 381-411.

Impact of Differential Injury Reporting on the Estimation of the Total Number of Work-Related Amputation Injuries

Sangwoo Tak, ScD, MPH, Kathleen Grattan, MPH, Lucy Bullock, BA, Letitia Davis, ScD, EdM, Leslie Boden, PhD, Al Ozonoff, PhD

Occupational Health Surveillance Program, Massachusetts Dept. of Public Health, Boston MA

ABSTRACT

Background
Capture-recapture methods have been used extensively to obtain estimates of illness or disease incidence and to assess coverage of surveillance systems. No studies have used this method to examine work related amputations. We aimed to estimate the total number of work-related amputations in Massachusetts that occurred in 2007 and 2008. Because amputations were defined based on a set of criteria, not all amputations were selected. We have examined the impact of this misclassification on the estimate of total number of work related amputations.

Methods
Eligible amputation cases include those with injury event dates during 2007 and 2008 and for which the event occurred while working. The Massachusetts Bureau of Labor Statistics (BLS) samples and workers' compensation records (WC) of lost time cases were used to perform a capture-recapture analysis. Records were linked using a specific set of matching criteria using FRIL® (v3.2). Case capture rates by data source were calculated. The total number of work related amputations was estimated using the Lincoln-Petersen (LP) estimator taking the BLS sampling weight into account. We found some of the unlinked cases were categorized as injuries other than amputations in either data source. We performed sensitivity analyses reassigning such cases as amputations.

Results
A total of 85 BLS sample and 381 WC cases were used in the analysis. A total of 22 BLS cases were found to be categorized as other injuries in WC data, and 23 WC cases were identified as other injuries in BLS. Depending on how we treat these cases, our estimates of the total number of work related amputations ranged from 276 to 442 cases which also yield dramatically different capture rates ranging from 35% to 87% for each data source.

Conclusion
Biased measures of work related amputations due to differential reporting would subsequently be misleading if used for prevention and evaluation purposes. Our findings highlight the importance of accurately classifying work related injuries and illnesses and caution researchers that potential misclassification of such cases can bias the estimates significantly and therefore they should do their due diligence in finding all the potential cases.

Exploring New Hampshire Workers' Compensation Data for its Utility in Enhancing the State's Occupational Health Surveillance System

Karla Armenti ScD§, Henry Vincent*, Rishika Nigam#, BS, MPH, and Alan Berko#
§New Hampshire Division of Public Health Services, Bureau of Public Health Statistics and Informatics, * New Hampshire Department of Labor, Workers' Compensation Division, # University of New Hampshire

Introduction

Workplace has an enormous impact on the health of the U.S. population. Nearly 3.1 million nonfatal workplace injuries and illnesses were reported among private industry employers in 2010, resulting in an incidence rate of 3.5 cases per 100 equivalent full-time workers. Workplace illnesses accounted for 5.1 percent of the 3.1 million injury and illness cases. [1]

Work-related injuries and illnesses impose a huge burden on workers, their families, businesses, and the economy. A new study, funded by the National Institute for Occupational Safety and Health (NIOSH), determined the cost of work-related injuries and illnesses in the United States to be $250 billion. This cost has risen by $33 billion since 1992, the last time a similar study was conducted. [2] In New Hampshire, workers' compensation claims alone cost approximately $239 million in 2008. [3]

Work-related injuries and illnesses can be prevented with appropriate and targeted interventions. Successful approaches to making the workplace safer begin with having the most accurate and current occupational health surveillance data, which are necessary to understand the root causes of the problems that lead to occupational injury and illness. [4] Unfortunately federal occupational health surveillance reporting requirements result in data gaps and shortfalls that do not accurately capture the true nature of occupational health and illness. This results in an inaccurate view that occupational health and illness is on a downward trend.

The major sources of occupational health data for surveillance purposes are: Bureau of Labor Statistics (BLS) Survey of Occupational Injury and Illnesses (SOII) and Census for Fatal Occupational Injury (CFOI), hospital discharge data and physician records, and state workers' compensation data. Data produced from these systems have been described as fragmentary, unreliable, and inconsistent, resulting in the underestimation of the true burden and magnitude of work related injuries and illnesses. [5]

The focus of this study was to better understand the contribution of workers' compensation data to surveillance of work-related

---

[1] United States Department of Labor, Bureau of Labor Statistics. Workplace Injury and Illness Summary, 2010, available at: http://bls.gov/news.release/osh.nr0.htm. Accessed July 17, 2012.
[2] Leigh, J. Paul, Economic Burden of Occupational Injury and Illness in the United States. The Milbank Quarterly, Vol. 89, No. 4, 2011 (pp. 728–772), http://onlinelibrary.wiley.com/doi/10.1111/j.1468-0009.2011.00648.x/pdf.
[3] Sengupta, I., Reno V, Burton JF., Workers Compensation: Benefits, Coverage, and Costs, 2008, September 2010, National Academy of Social Insurance.
[4] Friedman, L.S. and L. Forst, The impact of OSHA recordkeeping regulation changes on occupational injury and illness trends in the US: a time-series analysis. Occupational Environmental Medicine, 2007. 64(7): p. 454-60.
[5] Azaroff LS, Levenstem C, Wegman DH. "Occupational Injury and Illness Surveillance: Conceptual Filters Explain Underreporting." Am J Public Health. 2002 Sept; 92(Pt.9):1421-29.

injuries and illnesses in New Hampshire. We believe that WC data can be used for prevention priority setting purposes (as part of our fundamental, core occupational health surveillance program [6]); and to augment (not replace) what we know from other data sources, such as hospital discharge, death and cancer data, and labor statistics data.

While the focus of the WC system is primarily to set up and manage claims, and to submit required reports, there is valuable information that can be tracked to better understand the burden of work-related injury and illness on a state level. These include frequency of specific injuries, lost work time, severity, disability status, medical treatment and outcomes, types of injury, industry and establishment information, and cost. Any narrative text can also provide some data related to hazard identification.

Constraints do exist, however, that limit the usefulness of WC data for general public health surveillance purposes. Because WC data systems focus on management of the claims process and ratemaking (and are therefore insurance industry driven), there is less general health information and job related details. Illnesses are difficult to identify in WC data, due to the long latency period often associated with specific illnesses. Thus, acute injury data are more representative of worker population risks than are occupational illness data.

Although WC is an important data source for occupational health surveillance, not every state has unfettered access to record level data. There may be restrictive state privacy laws and confidentiality rules that can impede data sharing and analysis.

New Hampshire Workers' Compensation Data
Nearly 45,000 workplace injuries and illnesses are reported to the NH Department of Labor (NH DOL) in any given year. The Workers' Compensation Division of the New Hampshire Department of Labor was created in 1947 and has the responsibility for administration of the State's Workers' Compensation Law (RSA 281-A). This law originally enacted in 1911, requires employers to maintain insurance coverage to provide no fault workers' compensation for employees in case of accidental injury, death or occupational disease, "arising out of and in the course of employment" (RSA 281-A:2 XI).

Under the NH state law, every employer has to report to the Commissioner of the NH Department of Labor (DOL) any injury sustained by an employee in the course of employment as soon as possible, but no later than 5 days after the employer learns of the occurrence of such an injury (referred to as the First Report of Injury).

The law specifies the level of medical and wage replacement income benefit to be paid to injured workers and at the same time bars the employee from suing the employer for the injury. The division's coverage section is responsible for ensuring that all employers maintain this specific insurance coverage. The claims section's duties include scheduling and conducting hearings on contested cases, and monitoring the service of the insurance carriers to determine that benefit payments are provided in a timely manner. The Vocational Rehabilitation section is responsible for monitoring the vocational rehabilitation process. [7] The only exclusions in the law pertain to corporate officers and limited liability company members, who are not required to carry workers' compensation coverage on themselves until they have a fourth officer or member or a single employee. There is also no requirement for sole proprietors or for partners in partnerships to cover themselves for workers' compensation. [8]

---

[6] Occupational Health Indicators: A Guide for Tracking Work-Related Health Conditions and Their Determinants, at: http://www.cste.org/dnn/programsandactivities/occupationalhealth/occupationalhealthindicators/tabid/85/default.aspx
[7] State of New Hampshire, Department of Labor, 59th Biennial Report, July 1, 2009 – June 30, 2011. Available at http://www.labor.state.nh.us/biennialrpt.pdf.
[8] Title XXIII, Labor, Chapter 281-A, Workers' Compensation, Section 281-A:18-a, Exclusion of Executive Officers and Members of Limited Liability Companies.

There are many private carriers offering workers' compensation insurance in New Hampshire. Larger employers may be self-insured. Carriers are required to provide government agencies with claims information that is used for administrative purposes such as oversight, hearings for adjudication of disputes and other matters.

NH uses the first report of injury to establish a WC claim. The employer must file (or cause to be filed via the insurance company) the first report of injury to the NH DOL. The carrier has 21 days to investigate/review the claim and then accept or deny it. Employers can use an EDT –electronic data transfer to file the report.

The claim does not and will not contain any other medical information unless the claim is contested (denied) by the carrier. Of the 46,000 +/- total WC claims each year about 3-5% are contested. Contested claims include more medical treatment information as the hearing officer may need it to make his/her initial findings.

Methods

The NH Workers' Compensation data set was analyzed to better understand its potential in estimating work related injuries that occurred in 2008 and 2009, thereby enhancing the NH Occupational Health Surveillance System. This was a descriptive analysis of WC data, focusing on data completeness and data quality.

Data files in the NH WC system are arranged based on the date when the NH Department of Labor receives the work related injury or illness report irrespective of the date of injury or illness. We received two excel spreadsheets (2008 and 2009 data files) based on the year when DOL received the incident report. Twenty-eight indicators were present in each data file. Of importance for further analysis were Type of Injury, Cause of Injury, Type of loss, Treatment, Claim Type, Managed Care, Accident Premise.

Results

The total number of records received by the DOL in 2008 was 45,210 and in 2009 was 40,500 (based on date of report, not date of injury). In both 2008 and 2009 files, more than 40% of the data were missing from mostly all the field types except the Type and Cause of Injury. 41% of the data were missing from the Treatment field. Nearly 96% of the data were missing for Who Paid for the Medical Treatment. The data were filled completely for Age and Date of Injury but the age ranged from 0 to 99 years. For the 2008 WC data file, Date of Injury ranged from 1908 to 2008 and for 2009 WC data file, Date of Injury ranged from 1980 to 2009.

Table 1 presents the percent data missing from the 2008 and 2009 WC data files.

Type of Loss: For combined 2008 and 2009 data, 47% of the data were missing. 51% were Traumatic Injury, 1% were Cumulative Injury, and less than 1% were Occupational Disease. Treatment: For combined 2008 and 2009 data, 41% of the data were missing. 28% of the total number of workers received Minor Treatment at a clinic or hospital; 13% received No Treatment at all and 13 % received an Emergency Evaluation.

Accident Premises: For combined 2008 and 2009, 88% of data were missing. Of the 12% with Accident Premises information, 77% were Employer's Location and 74% in 22% were listed as Other.

Discussion

Workers' compensation data are a valuable source for documenting work-related injuries and illnesses. This information can improve our understanding of the causes and prevention of workplace injury and disease. Missing data, however, is something that can impact interpretation and must be considered when utilizing workers' compensation data for population based surveillance. The amount of missing fields in the WC data set makes it difficult to determine the true incidence of occupational injury and illness. Using WC data in conjunction with other public health data sets provides additional important information in tracking occupational injuries and illnesses in the United States.

Table 1. Percent data missing from the WC data files. (File names are based on the year when NH DOL received the injury report)

Indicators	Data missing in percent (%)	
	2008 WC data file (n=45210)	2009 WC data file (n=40500)
Injury cause	72.00	75.68
Injury extent	72.01	75.68
Injury body	72.00	75.68
Injury date	0.31	0.14
SIC industry code	67.73	55.86
NAICS industry code	56.49	56.64
WCIO injury code	12.10	10.80
WCIO body part code	12.03	10.74
WCIO injury cause	12.08	10.78
Initial treatment code	41.29	41.16
Type of loss	47.64	46.12
Accident premise	87.47	87.89

Figure 1. Percentage of Nature and Cause of Injury

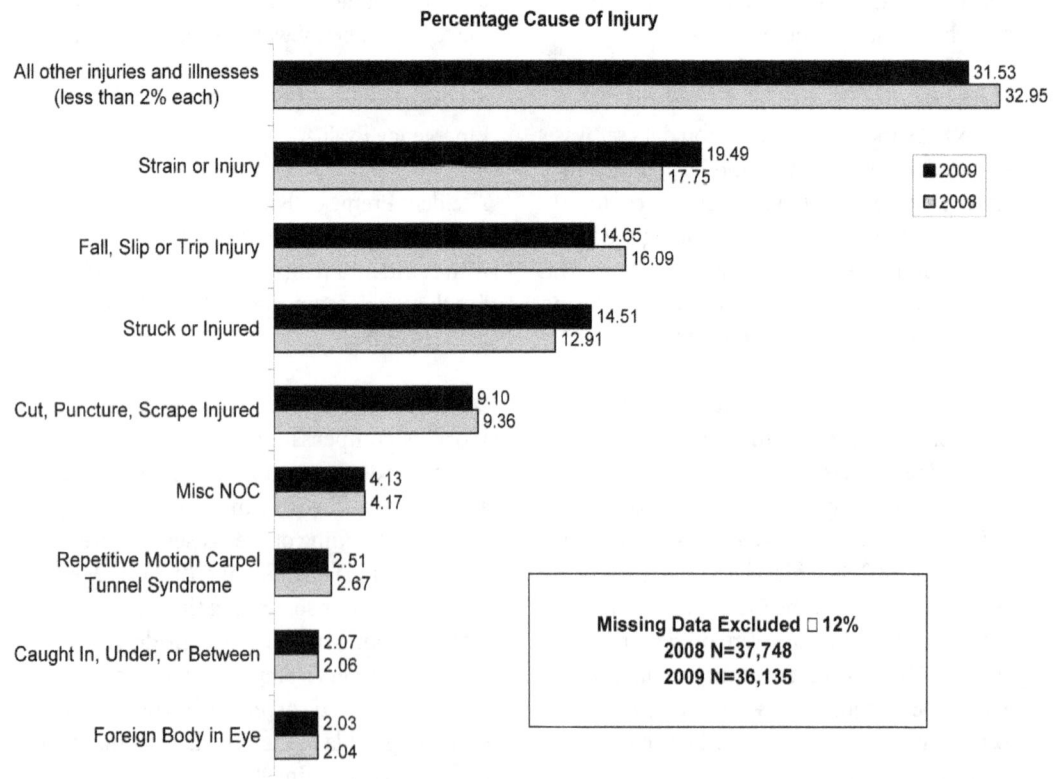

Figure 2. Percentage Nature of Injury

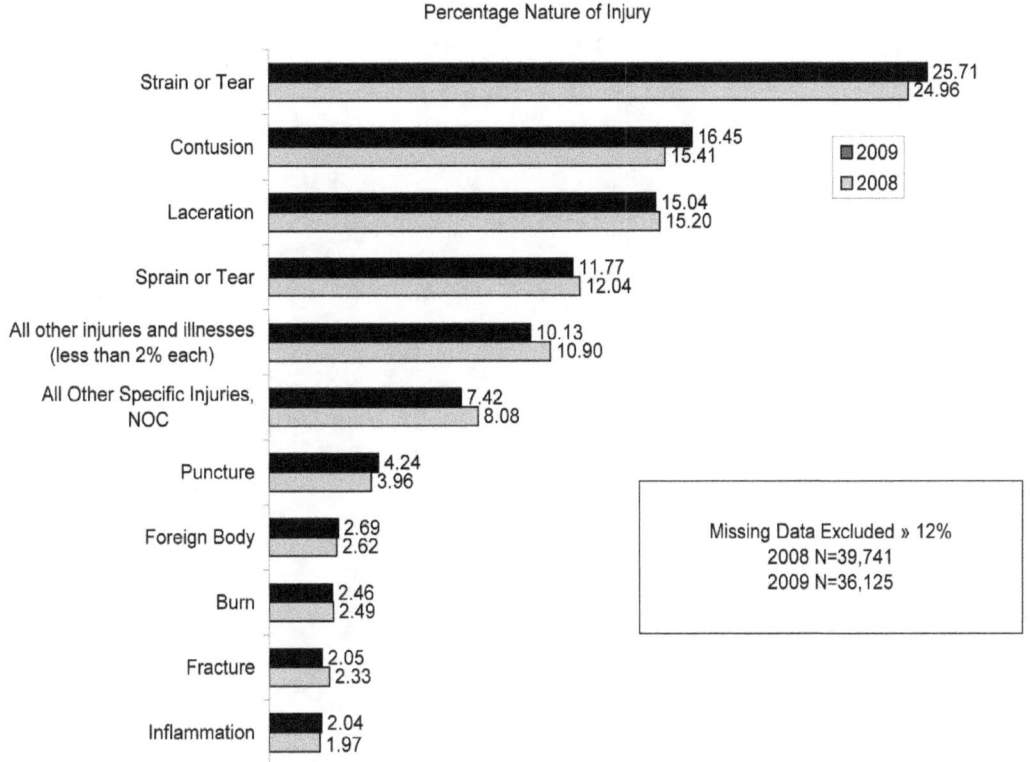

Using Workers' Compensation Data for Surveillance of Occupational Injuries and Illnesses – Ohio, 2005–2009[1]

Alysha Meyers, PhD§, Steve Wurzelbacher§, PhD, Steve Bertke§, PhD Mike Lampl*, MS, Dave Robins*, Jennifer Bell§, PhD

§National Institute for Occupational Safety and Health, *Ohio Bureau of Workers' Compensation, Division of Safety & Hygiene

Introduction

The Ohio Bureau of Workers' Compensation (OBWC) is the largest of four, exclusive, state-run workers' compensation (WC) systems in the United States. All public Ohio employers and private employers (except sole proprietorships or partnerships) with less than 500 employees must purchase WC insurance from the state of Ohio. Other private employers have the option to self-insure for WC insurance. OBWC provides WC insurance for approximately two-thirds of Ohio workers but a lower proportion of Wholesale and Retail Trade (WRT) industry sector workers. One long-term goal of a partnership between OBWC and the National Institute for Occupational Safety and Health (NIOSH) is to develop an occupational illness and injury surveillance system by joining OBWC's WC data with denominator data (number of employees) from another state agency. To demonstrate the new system NIOSH generated incidence rates for OBWC WC outcomes, especially for musculoskeletal disorders (MSDs), for single-location employers in the Wholesale and Retail Trade Sectors and described sub-sector trends for 2005–2009. MSDs caused by ergonomic hazards are common among workers and result in pain, disability, and substantial cost to workers and employers (Bureau of Labor Statistics, 2011; Liberty Mutual Research Institute for Safety, 2011). Based on data from the 2010 Bureau of Labor Statistics (BLS) Survey of Occupational Injuries and Illnesses (SOII), a disproportionately higher rate of MSDs resulting in lost workdays occurs in the WRT industry sector compared with other industry sectors (Bureau of Labor Statistics, 2011).

Methods

Claims data for this system were extracted from the OBWC data warehouse; denominator data (number of employees) and North American Industry Classification System (NAICS) codes came from the Ohio Department of Jobs and Family Services (ODJFS). Joining the databases was complex. ODJFS data are arranged by employer location, whereas OBWC data are arranged by policy number. OBWC claims cannot be linked to a particular location if the policy includes more than one location. Therefore, these analyses focused on single-location employers in the WRT industry sector. Among OBWC-insured policies, the vast majority are for employers with a single-location. (2009 OBWC-insured WRT policies: single-location — 31,599 and multiple-location — 882). To join single-location denominator data to single-location OBWC policy data, NIOSH and OBWC first developed a method to identify active policies by year. ODJFS' quarterly data were annualized to calculate the average number of employees per employer. Single-location ODJFS data were joined to OBWC policy data by year (2005–2009) using Employer Identification Numbers (i.e. Federal Tax Identification Numbers) common to both databases. Rarely (< 1% of policies), more than one OBWC policy matched to one ODJFS employer; in those cases the employer's data were excluded.

---

[1] The findings and conclusions in this paper are those of the authors and do not necessarily represent the views of the National Institute for Occupational Safety and Health or the Ohio Bureau of Workers' Compensation.

With few exceptions, MSDs were defined according to the BLS case definition. Coded injury/illness diagnosis data and narrative text on causation were used to identify MSD claims; a Bayesian auto-coding technique (Lehto et al., 2009) used both data elements to identify MSDs by using a 'training' and 'testing' set of manually coded claims. The sensitivity and specificity of this auto-coding technique were 0.90 and 0.98 respectively. Auto-coded MSD claims were flagged for manual, expert review when the injury/illness diagnosis was not an MSD. Lost-time claims for MSDs were defined as claims for MSDs resulting in more than seven days away from work. Rates of MSDs were calculated per 10,000 employees and tests of trends over time were calculated using Poisson regression. Disallowed and dismissed claims were excluded from all analyses.

Results
The proportion of all claims attributable to MSDs was relatively stable at approximately 20% across the 5-year period of 2005–2009; the proportion of MSD lost-time claims decreased from 37% in 2005 to 32% in 2009. From 2005–2009, the majority of claimants were men, 25-54 years of age, and worked for employers with 11-249 employees. The largest number of MSD claims occurred among Merchant Wholesalers of Durable Goods.

The rate of MSDs resulting in a claim or a lost-time claim decreased significantly (P < .05) from 2005–2009 for all WRT industry sector employers but not for all WRT industry subsectors. Overall, the respective rates of all MSD claims and lost-time MSD claims per 10,000 employees decreased from 86.3 and 28.7 in 2005 to 52.8 and 14.1 in 2009. Employers with more employees tended to have higher incidence rates of MSDs. From 2005–2009 lost-time MSD rates per 10,000 employees for three subsectors were in the highest five every year: Merchant Wholesalers of Nondurable Goods (2009: 29.2), Furniture and Home Furnishings Stores (2009: 21.7), and Merchant Wholesalers of Durable Goods (2009: 15.5) (Figure 1). The high lost-time MSD rates per 10,000 employees in these three subsectors were consistently attributable to high rates in five 4-digit NAICS industry subsector groups: a) in Merchant Wholesalers of Nondurable Goods high rates were attributable to Alcoholic Beverage Merchant Wholesalers (2009: 114.8) and Grocery and Related Product Wholesalers (2009: 30.9); b) in Furniture and Home Furnishings Stores high rates were attributable to Furniture Stores (2009: 27.2); and c) in Merchant Wholesalers of Durable Goods high rates were attributable to Metal and Mineral Merchant Wholesalers (2009: 28.0) and Motor Vehicle and Motor Vehicle Parts and Supplies Merchant Wholesalers (2009: 25.4).

Discussion
Improved surveillance of work-related MSDs is a national priority identified in NIOSH National Occupational Research Agenda (National Institute for Occupational Safety and Health (NIOSH) Office of the Director, 2012). This project demonstrates how WC claims data can be used for public health surveillance. The results indicate that while the rate of MSDs among workers employed by smaller (< 500 employees) Ohio employers declined from 2005 to 2009 for most WRT industry subsectors, workers in some subsectors are still experiencing relatively high rates of MSDs. The factors responsible for the downward trend in incident MSDs in most WRT industry subsectors are unclear.

A downward trend for incident MSDs from 2005–2009 was also observed in the BLS SOII (Bureau of Labor Statistics, 2011). Based on OBWC data, the WRT industry subsector with the highest rate of lost-time MSDs for five straight years (2005–2009) was Merchant Wholesalers of Nondurable Goods. Similarly, Merchant Wholesalers of Nondurable Goods had one of the three highest rates of MSDs among all WRT industry subsectors over the same 5-year period in the BLS survey (Bureau of Labor Statistics, 2011). In both cases these high rates were attributable to high rates among Alcoholic Beverage Merchant Wholesalers and Grocery and Related Product Wholesalers. Based on OBWC data, from 2005–2009 Furniture and Home Furnishing Stores had one of the three highest rates of MSDs for three years and one of the five highest rates for all years. Similarly, in the BLS survey, Furniture and Home Furnishing Stores had one of the five highest rates of MSDs for the same 5-year period (Bureau of Labor Statistics, 2011). From 2005–2009 Building Material and Garden Equipment and Supplies Dealers had one of the three highest rates of MSDs based on the BLS survey and during the same time period that subsector was among the five highest OBWC lost-time MSD rates for four of five years. In contrast, from 2005–2009, Food and Beverage Stores had one of the five highest rates of MSDs in the BLS survey but were never among the five highest subsectors among OBWC data. Given the differences between the BLS survey and the OBWC claims data, it is not surprising that the relative MSD rates may vary. Cases in the BLS survey results are from OSHA logs and describe injuries/illnesses for employers of all sizes at one location among workers who missed at least one day of work. In contrast, OBWC cases are all for WC claims and describe injuries/illnesses for small, single-location employers among workers who did and did not miss work. Also, the majority of Food and Beverage Stores' employers may be chains that tend to self-insure and/or have multiple-locations and therefore these employers would not be represented by this project.

Workers in the higher risk subsectors of WRT are exposed to physical risk factors for MSDs such as overexertion or repetitive motion (Anderson et al., 2010). Work tasks in high risk subsectors such as Furniture and Home Furnishings Stores, Alcoholic Beverage Merchant Wholesalers, and Grocery and Related Product Wholesalers commonly include lifting and transporting large heavy objects such as furniture or kegs of beer. OSHA has created ergonomic training tools that outline prevention activities for Beverage Delivery and Grocery Warehousing (OSHA, 2012). Some interventions (e.g. stair-climbing dollies, keg handling equipment, forklifts) to reduce exposures exist for many but not all manual material handling tasks in these subsectors.

The findings from this project are subject to at least three limitations. First, this project is only representative of smaller employers (<500 employees) with a single location in Ohio. Second, the Bayesian auto-coding method used to identify MSDs introduces the potential for non-differential misclassification. However, misclassification is not expected to create bias in MSD rates by WRT industry subsector. Third, underreporting is an expected limitation of WC data. Studies have estimated that WC claims data underreports work-related injuries/illnesses by 40–80% (Weddle, 1996; Pransky et al., 1999; Morse et al., 2000). While underreporting may affect the magnitude of the rates, it is unknown whether the relative differences observed between WRT industry subsector or employer sizes were affected by underreporting.

The findings from this project suggest that the incidence of MSDs has declined from 2005 to 2009 among small WRT employers in Ohio. The data also indicate that relatively higher rates of MSDs occur in the Alcoholic Beverage Merchant Wholesalers, Grocery and Related Product Wholesalers, Furniture Stores, Metal and Mineral Merchant Wholesalers, and Motor Vehicle and Motor Vehicle Parts and Supplies Merchant Wholesalers industry subsector groups. The WRT industry subsector findings are consistent with national BLS surveillance data. Targeted interventions to reduce exposure to ergonomic hazards in these subsectors should continue to be developed and implemented to effectively prevent MSDs. Given the large workforce employed in WRT industry sector, declines in MSDs could substantially reduce the national injury/illness burden.

References
Anderson, V. P., P. A. Schulte, et al. (2010). "Occupational Fatalities, Injuries, Illnesses, and Related Economic Loss in the Wholesale and Retail Trade Sector." *American Journal of Industrial Medicine* 53(7): 673-685.

Bureau of Labor Statistics (2011). Nonfatal occupational injuries and illnesses requiring days away from work, 2010. Bureau of Labor Statistics News Release, U.S. Department of Labor.

Bureau of Labor Statistics. (2011, December 8). "Survey of occupational injuries and illnesses. Nonfatal (OSHA recordable) injuries and illnesses. Case and demographic characteristics for work-related injuries and illnesses involving days away from work. Special tabulation — Incidence rate and number of nonfatal occupational injuries and illnesses involving days away from work by selected industries with musculoskeletal disorders, 2003–2010." from http://www.bls.gov/iif/oshcdnew.htm.

Lehto, M., H. Marucci-Wellman, et al. (2009). "Bayesian methods: a useful tool for classifying injury narratives into cause groups." *Injury Prevention* 15(4): 259-265.

Liberty Mutual Research Institute for Safety (2011). 2011 Liberty Mutual Workplace Safety Index. Hopkinton, MA: 2.

Morse, T., C. Dillon, et al. (2000). "Reporting of work-related musculoskeletal disorder (MSD) to workers compensation." *New Solutions: a journal of environmental and occupational health policy* 10(3): 281-292.

National Institute for Occupational Safety and Health (NIOSH) Office of the Director. (2012, February 8, 2012). "The National Occupational Research Agenda (NORA),." Retrieved February 16, 2012, from http://www.cdc.gov/niosh/nora/.

OSHA. (2012). "Ergonomic eTools." eTools, eMatrix, Expert Advisors and v-Tools Retrieved April 4, 2012, from http://www.osha.gov/dts/osta/oshasoft/index.html.

Pransky, G., T. Snyder, et al. (1999). "Under-reporting of work-related disorders in the workplace: a case study and review of the literature." *Ergonomics* 42(1): 171-182.

Weddle, M. G. (1996). "Reporting occupational injuries: The first step." *Journal of Safety Research* 27(4): 217-223.

Figure 1. Rates of lost-time musculoskeletal disorders per 10,000 employees among Wholesale and Retail Trade industry subsectors

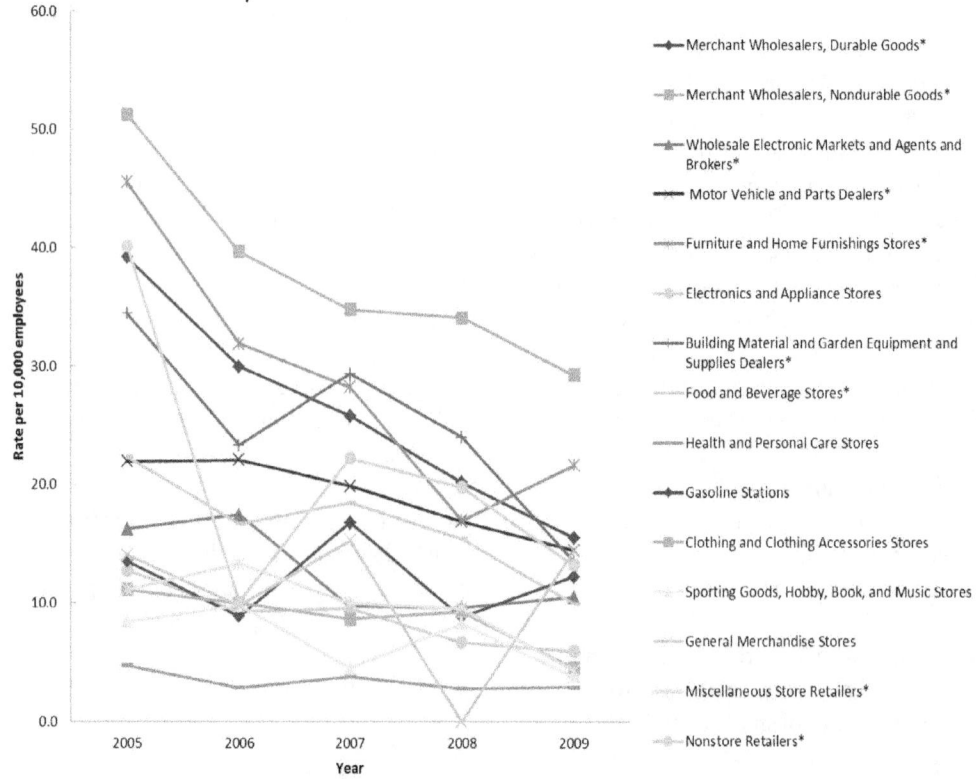

Using an Administrative Workers' Compensation Claims Database for Occupational Health Surveillance in California: Validation of a Case Classification Scheme for Amputations

Rachel Roisman§, Lauren Joe*, Matthew Frederick#, Stella Beckman*, John Beckman#, Martha Jones†, Robert Harrison§

§ Occupational Health Branch, California Department of Public Health, *Centers for Disease Control and Prevention/Council of State and Territorial Epidemiologists Applied Epidemiology Fellow, California Department of Public Health, #Impact Assessment Inc., †Division of Workers' Compensation, California Department of Industrial Relations

Introduction

Amputations can be severe injuries that result in a worker being unable to perform his or her original job and having reduced earning potential and/or permanent disability. The most common work-related amputation involves a finger but also may involve larger body parts such as a hand, foot, arm, or leg. The major source of data on the number of amputations in California comes from the U.S. Bureau of Labor Statistics (BLS) Annual Survey of Occupational Injuries and Illnesses (SOII). The BLS estimated 10,000-12,000 non-fatal, work-related amputations per year (1992-1999) in the United States (Brown 2003). The SOII estimate is based on a probability sample of employer reports of injuries and illnesses resulting in at least one day away from work and recorded on an OSHA-300 log. For 2007 and 2008, the SOII estimated a total of 1,390 work-related amputations in California (BLS 2007, 2008). Work-related amputations are preventable, and public health surveillance can track these injuries, estimate the burden, identify high-risk industries and occupations, and guide prevention activities and regulatory changes to better protect workers. Exploring additional data sources for information about work-related amputations in California is critical to ensuring an accurate enumeration of the total public health burden of amputations.

In 2000, the California Department of Industrial Relations (DIR) Division of Workers' Compensation (DWC) implemented the electronic Workers' Compensation Information System (WCIS). California uses a standardized format for the WCIS as outlined by the International Association of Industrial Accident Boards and Commissions. Claims administrators must submit an electronic First Report of Injury to the WCIS within five working days after knowledge of an injury or illness. Additional data are submitted to the system any time other action is taken on a claim, as well as annually on all claims. The electronic WCIS in California contains about nine million records (2000-present) and is one of the largest repositories of detailed workplace injury and illness data in the United States.

WCIS was created for administrative purposes (i.e., tracking information related to individual workers' claims, medical services, and related payments) but can be used for occupational health surveillance. The California Department of Public Health Occupational Health Branch (CDPH-OHB) has used WCIS for public health surveillance of work-related asthma and pesticide illness for several years. Previous work done by CDPH-OHB using WCIS to enumerate amputations in California used the Nature of Injury (NOI) code, included claims that resulted in lost workdays, and excluded claims where the Part of Body (POB) was thought to be unlikely to be associated with an amputation (e.g. eye, back, chest) (CDPH, 2012); this analysis counted a total of 1,736 work-related amputations in California in 2007 and 2008.

In 2010, CDPH-OHB received funding from the BLS to utilize WCIS, in addition to two other data sources, for surveillance of work-related amputations and carpal tunnel syndrome. To investigate whether NOI is an accurate indicator of a true amputation, and to lay the groundwork for a better understanding of the burden of work-related amputations in California, we reviewed a sample of cases that occurred in 2007-2008, developed a case classification scheme, reviewed medical records to validate the scheme, and calculated positive predictive value (PPV) and negative predictive value (NPV). These data can be used to enumerate work-related amputations in California for which a claim was filed.

Methods

We requested a data extract from DIR containing information on the injury, employee, employer, and benefit payments for all claims in WCIS with an injury or claim date between 2007 and 2008. Initial case definitions were developed in conjunction with the BLS, DWC, the Massachusetts Department of Public Health, Washington State Department of Labor and Industries, and Boston University (all partners in the BLS-funded project). Using the amputation diagnosis and procedure codes identified by the group, we added procedure and diagnosis data from the medical billing data for claims identified as potential amputations. This original amputation extract from the WCIS database included all claims with a NOI of "amputation," appropriate keywords in the Injury Description (ID), or with appropriate diagnosis or procedure codes in the medical billing data. This extract contained 6,605 claims for 2007-2008.

Manual review of a sample of amputation claims revealed that the initial extract contained claims that were not amputations. Therefore, a classification scheme was established by reviewing cases and examining characteristics that suggested the level of probability that an extracted case was a true amputation case. Amputation cases were classified as probable or uncertain based on a combination of values in the following fields: diagnosis, procedure, NOI, POB, cause of injury, and ID. Cases with amputation diagnosis, amputation-related procedure codes, or a NOI signifying amputation were considered probable amputations. Both POB and ID were used as "criteria variables" to support probable codes and the NOI code. A probable case had a [probable code and/or NOI = amputation] and a [strong ID and/or acceptable POB]. An uncertain case did not have this combination of variables. Table 1 delineates all the possible combinations of the variables and the resulting case classification.

The case classification scheme was confirmed on a preliminary basis by further review of classified cases to ensure that the majority of cases in a given class appeared to be properly classified; examples of actual cases are shown in Table 2. The case classification was validated by medical record review of select amputation claims. A random sample of 100 amputation medical records was requested. Complete medical records were reviewed by two physicians independently, and positive and negative predictive values (PPV, NPV) were calculated.

Results

Of the 100 medical records requested, 75% were classified as probable amputations and 25% were classified as uncertain amputations; these percentages reflected the breakdown of cases in the extract itself (Table 3). Fifty-three medical records (72% probable and 28% uncertain) were received and reviewed (Table 3). There was

---

[1] NOI codes follow Workers' Compensation Insurance Organizations (WCIO) codes; "02" corresponds to "amputation."
[2] Injury Description keywords: "bony loss," "cut off," "amputation," or variation.
[3] 5 Diagnosis Related Group (DRG) codes, 68 International Classification of Diseases, Ninth Revision, Clinical Modification (ICD-9-CM) diagnosis codes
[4] 95 Current Procedural Terminology (CPT) codes; 46 ICD-9 CM procedure codes; 124 Healthcare Common Procedure Coding System (HCPCS) codes.

Table 1. Amputation Case Classification

Probable Code[5]	Nature of Injury = Amputation	"Strong" Injury Description[6]	Acceptable Part of Body[7]	Case Classification
Yes	Yes	Yes	Yes	Probable
Yes	Yes	Yes	No	Probable
Yes	Yes	No	Yes	Probable
Yes	Yes	No	No	Uncertain
Yes	No	Yes	Yes	Probable
Yes	No	Yes	No	Probable
Yes	No	No	Yes	Probable
Yes	No	No	No	Uncertain
No	Yes	Yes	Yes	Probable
No	Yes	Yes	No	Probable
No	Yes	No	Yes	Probable
No	Yes	No	No	Uncertain

Table 2. Amputation Case Examples

Case Class	Probable Amputation	Uncertain Amputation
Diagnosis or Procedure Code	ICD-9, 866.0 (Amp of Finger)	ICD-9, 724.2 (Lumbago)
Nature of Injury	Crushing	Amputation
Injury Description	SMASHED RT MID FGR DYE CAME DOWN ON HARD	HE COMPLAINS THAT HIS LOWER BACK IS BOTHERING HIM
Part of Body	Finger(s)	Low back area

Table 3. Amputation Medical Record Review

Case Class	2007-2008 Amputation Extract (n = 6,605)		Medical Records Requested (n = 100)		Medical Records Received (n = 53)	
	n	%	N	%	n	%
Probable	5,097	77.2	75	75.0	38	71.7
Uncertain	1,508	22.8	25	25.0	15	28.3

---

[5] Amputation Diagnosis: ICD-9-CM code of 885, 885.0, 885.1, 886, 886.0, 886.1, 887, 887.0, 887.1, 887.2, 887.3, 887.4, 887.5, 887.6, 887.7, 895, 895.0, 895.1, 896, 896.0, 896.1, 896.2, 896.3, 897, 897.0, 897.1, 897.2, 897.3, 897.4, 897.5, 897.6, 897.7, V4970, V4960
Amputation Procedure: CPT code of 23900, 23920, 23921, 24900, 24920, 24925, 24930, 24931, 24940, 25900, 25905, 25907, 25909, 25915, 25920, 25922, 25924, 25927, 25929, 25931, 26910, 26951, 26952, 27290, 27295, 27590, 27591, 27592, 27594, 27596, 27598, 28800, 28805, 28810, 28820, 28825, 11752, 54120, 54135, 69110, 69120

[6] Contains "amputate" or contains "severed" (without "tendon")

[7] Acceptable Part of Body (numeric codes are WCIO codes): 30 - Multiple Upper Extremities; 31 - Upper arm; 32 – Elbow; 33 - Lower Arm; 34 – Wrist; 35 – Hand; 36 - Finger(s); 37 – Thumb; 38 - Shoulder(s); 39 - Wrist(s) & Hand(s); 50 - Multiple Lower Extremities; 52 - Upper Leg; 53 – Knee; 54 - Lower Leg; 55 – Ankle; 56 – Foot; 57 - Toe(s); 58 - Great Toe; 90 - Multiple Body Parts.

Table 4. Case Classification Validation

Case Class in WCIS	Case Class Determined by Reviewers		Total
	Probable	Uncertain	
Probable	31 *True positives*	7 *False positives*	38
Uncertain	0 *False negatives*	15 *True negatives*	15
Total	31	22	53

complete concordance by the two physicians who independently reviewed the medical records.

Of the 53 cases reviewed, 13% were misclassified by our case classification (Table 4). All of the misclassified cases were false positives, resulting in a PPV of 0.82 and a NPV of 1.0. Based on this sample, a case classification of "probable" is correct 82% of the time, while a case classification of "uncertain" is correct 100% of the time. Results suggest that our case classification overestimates the actual number of probable amputations by 20%, while uncertains are likely to be classified correctly. Therefore, we determined that "uncertain" amputation cases should not be included in a final enumeration of amputations in WCIS, which reduced the number of amputations counted in 2007-2008 from 6,605 to 5,097.

Discussion

This represents the first time, to our knowledge, that an amputation case classification scheme has been developed for the California WCIS. Using this method, we estimated a total of 5,097 work-related amputations in California in 2007-2008; this is 3.7 times the number of amputations estimated by SOII. Recent studies have suggested that the SOII underestimates the number of occupational injuries and illnesses (Rosenman et al. 2006; Boden et al. 2008a; Boden et al. 2008b). However, there are a number of reasons why SOII and WCIS numbers are not directly comparable. For example, SOII and WCIS do not use the same inclusion criteria (e.g., SOII only includes cases with one or more days away from work, whereas we included all claims regardless of lost work time). In addition, due to incomplete information on claim result, we included both accepted and denied claims in our analysis of WCIS data, whereas employers may only include cases for SOII reporting that have been deemed work-related.

Inspection of individual claims revealed that diagnosis, procedure, and NOI codes alone are not enough to identify amputation claims. The reasons for this discrepancy are not fully known. Data entry error certainly could be a component. Several limitations to our approach may also contribute to the discrepancy: although clinically an amputation includes bone loss, we were not able to verify whether bone loss occurred as this was not necessarily noted in WCIS. Therefore, we likely included amputations that would not clinically be considered an amputation. There is a question as to how well secondary amputations or those occurring after the initial recorded injury are captured by the system. For instance, an initial crush injury that resulted in an amputation several weeks or months later may or may not be considered an amputation using this case classification.

While WCIS is an administrative dataset, we have demonstrated that it can be used for public health surveillance. The WCIS contains detailed case data that can be used for targeting high-risk occupations and industries. Although this analysis does not address the issue of undercounting directly, it does suggest that neither the SOII, nor relying on NOI codes in WCIS alone, captures the full public health burden of work-related amputations in California. Even our estimate is expected to be an undercount since some workers are not covered by workers' compensation [8], and workers who are eligible may not file for workers' compensation for a variety of reasons (Azaroff 2002). Significant work has gone into completely and accurately identifying amputation cases in WCIS, and these methods can be used in the future to enumerate cases of this condition in years other than 2007 and 2008. If additional endpoints are to be enumerated, resources are necessary to develop a new case classification scheme for each endpoint of interest. However, for targeted intervention activities driven by a priori knowledge of feasible prevention efforts, the effort required to develop a case classification scheme for WCIS may be worthwhile as it will likely substantially improve the enumeration of the problem and provide enhanced assurance of the accuracy of the enumeration.

References

Azaroff LS, Levenstein C, and Wegman DH. [2002] Occupational injury and illness surveillance: Conceptual filters explain underreporting, *Am J Public Health* 92:1421-1429.

Boden LI and Ozonoff A. [2008a] Capture-recapture estimates of nonfatal workplace injuries and illnesses, *Annals of Epidemiology* June:500–06.

Boden LI and Ozonoff A. [2008b] Reporting Workers' Compensation Injuries in California: How Many are Missed, The California Commission on Health and Safety and Workers' Compensation, August. Available at: http://www.dir.ca.gov/chswc/reports/reportingworkerscompensationinjuriesincalifornia2008august.pdf Accessed July 19, 2012.

Brown JD. [2003] Amputations: A Continuing Workplace Hazard. Bureau of Labor Statistics. Originally Posted: January 30, 2003. Available at: http://www.bls.gov/opub/cwc/sh20030114ar01p1.htm#7 Accessed July 19, 2012.

California Department of Public Health – Occupational Health Branch [2012]. California Occupational Health Indicators Report: Annual measures of worker health and safety for years 2003 – 2008, June. Available at: http://www.cdph.ca.gov/programs/ohsep/documents/allindicators03-08.pdf Accessed July 19, 2012.

Rosenman KD, Kalush A, Reilly MJ, Gardiner JC, Reeves M, and Luo Z. [2006] How much work-related injury and illness is missed by the current national surveillance system? Journal of Occupational and Environmental Medicine April:357–65.

U.S. Bureau of Labor Statistics Annual Survey of Occupational Injuries and Illnesses. Available at: http://www.bls.gov/respondents/iif/ Accessed July 19, 2012.

---

[8] Workers' compensation coverage and electronic reporting is mandatory for all California employers with at least one employee, except for those covered by the Federal Employees' Compensation Act, the Longshore and Harbor Workers' Compensation Act, the Federal Employers' Liability Act, or the Jones Act. In addition, coverage does not extend to those employed by American Indian tribes who have not waived their sovereign immunity.

Describing Agricultural Occupational Injury in Ohio Using Bureau of Workers' Compensation Claims[1]

Jedidiah A. Bookman§, David Robins*, Mamta Mujumdar*, S. Dee Jepsen§
§The Ohio State University, Department of Food, Agricultural and Biological Engineering. *Ohio Bureau of Workers' Compensation, Division of Safety and Hygiene

Introduction

Agriculture is one of the most dangerous industries in the United States (U.S. Department of Labor, 2010). Although the dangers of production agriculture are high, little concrete evidence exists to explain the extent of injury and the types of injuries sustained to Ohio agricultural producers. In Ohio, agriculture is attributed to employment for one of every seven Ohioans (Ohio Department of Agriculture, 2010) and is comprised by approximately 75,800 farms (United States Department of Agriculture - National Agricultural Statistics Service, 2009). Due to a high number of agricultural workers in Ohio, efforts are needed both to quantify injury data and prevent injuries in this vital portion of Ohio's economy.

Bureau of workers' compensation claims are a viable source of injury data and have been used in other states to quantify agricultural occupational injury. Similarities exist in previous studies and an exploration of agricultural claims collected by the Ohio Bureau of Workers' Compensation (Ohio BWC). Monopolistic workers' compensation systems, like that of Ohio, are great areas of opportunity in research. Having one provider of insurance in the state can greatly increase the probability of capture of an injury.

Although past studies have identified agricultural injury using workers' compensation data, this is the first effort in Ohio. Also unique to this study, injury claims were collected based on injury event coding and not primary risk category code of the employer. This paper explains the process and findings that investigated agriculture related injuries in Ohio. The study objectives were: (1) determine the number, cost, severity, and type of agriculture related injuries, (2) ascertain injury rates for the workers covered in the study, and (3) explore the utility of the Ohio Bureau of Workers' Compensation's database as a valid source of agricultural injury data.

Methods

Data Sample
All injury claim data was supplied by Ohio BWC for years 1999-2008. Data for the claims are derived from First Report of Injury forms and additional follow-up between claims representatives and injured workers. Fields included with each claim contained 30-month costs, International Classification of Disease version 9 (ICD-9) codes for most limiting injury, worker age at injury, and worker time away from work. Categorical codes were included on an industry level using the NCCI manual coding system for each claim. At the specific injury level, an NCCI manual code was assigned by a claims specialist after reviewing evidence on each claim. Starting in 2008, causation topics were collected with claim reporting. Causation was split between type of injury and object causing the injury.

Data Collection
The total claim pool was searched based on

---

[1] The findings and conclusions in this paper are those of the authors and do not necessarily represent the views of the Ohio Bureau of Workers' Compensation

15 occupational classifications established by the Ohio BWC (Table 1) (Ohio Bureau of Workers' Compensation, 2011) that are related to the North American Industry Classification System (NAICS) definition for 'Agriculture'.

Limited data were available at the industry level in the data. Because of these limitations, all claims were searched based on injury event coding. Using this method, all injuries involving agricultural production would be captured, regardless of primary industry of the employer. Due to confidentiality concerns, the complete dataset was housed within Ohio BWC and only aggregated data were released to the researcher to be published.

Claim Validity

The initial data query yielded 18,688 claims from the occupational groups listed in Table 1. An initial review of the results prompted further exploration to assess validity of the injury claims. A large proportion of claims seemed misclassified from reviewing 'injury text', 'firm name', and 'occupation name' fields. For example, temporary employment leasing agencies were represented in the initial data set. All claims by the agencies had the same injury level coding regardless of worker's assigned task or employment location. It was apparent errors could exist in the coding of the injury claims. A systematic review of each claim was then conducted to discern claim relation to agricultural production. The reviewer used 'injury text', 'firm name', 'city', 'county', 'occupation name', and 'primary name' to conduct an internet search of the firm. If results yielded any data related to agricultural production, the claim was deemed valid. Also, if injury text or occupation name mentioned any keywords related to agricultural production, the claim was considered valid. After screening for validity, the size of the claim pool was decreased by 24 percent (n=14,344) (Figure 1).

Data Analysis

Inter-rater reliability tests were performed to ensure validity of the review process. Minitab statistical package was used to calculate the Fleiss' Kappa value for inter-rater agreement between the researcher and three Ohio BWC Division of Safety and Hygiene professionals. The researcher's judgment of claim validity, either "yes" (valid) or "no" (invalid) was used as the benchmark to which the other three raters were compared.

An a priori confidence interval of 95% and agreement level of K=0.75 were selected based on literature review (Fleiss, Levin, & Paik, 2004). Results showed excellent agreement beyond chance for the three raters and the researcher's scoring of injury claims to be valid in terms of actually related to agriculture. Descriptive analyses were performed for total costs of injury, days lost, age, and gender. Using the ICD-9 code for each injury claim, body location and injury type was categorized. Of the 14,344 claims examined, 99.5% had attached optimal return to work ICD-9 codes (n=14,287). The optimal return to work code categorizes the injury that most inhibits the worker from returning to work. For example, if a worker has a broken arm and a contusion, the code would be created for the broken arm, as it will take longer to heal and would prevent the worker from returning to the job longer. Using the Barell Body Region by Nature of Injury Diagnosis Matrix (Barell et al., 2002), body region and nature of injury was established for 89.3% of total claims examined (n=12,814). An estimate of the total number of agricultural workers at-risk of injury was created using employer reported annual payroll and the average weekly wage of agricultural workers in the state found from a review of literature (Demers & Rosenstock, 1991; Villarejo, 1998). Approximately 50% of the 14,344 injury claims (n=7,054) had supplied wage data between the 2001-2008 years. The formula used to derive approximate employee counts for an individual firm is shown below:

$$(Employment\ per\ Year) = \frac{\left[\frac{Reported\ Yearly\ Payroll}{State\ Average\ Weekly\ Wage}\right]}{52\ Weeks\ per\ Year}$$

Results

It is important to note the differences between the industry and claim level of coding for the injury claims. Each claim has a unique claim level code based on the workers' occupation collected at the time of the injury report. Some industry and claim level classification categories matched, but the majority did not. Industry level of coding is derived from an algorithm investigating the proportion of payroll devoted to different occupations within the firm. Any claim analysis based on industry level will be coded by the type of industry the employer belonged and not the type of occupation performed by the injured worker.

A total of 14,344 claims were analyzed. A decreasing trend was observed in both total injuries and injury rate. Table 2 shows total costs for each occupation group as well as average cost per claim. Figure 2 shows the rate of injury for all agricultural workers captured in the dataset for years 2001-2008 and the total injuries observed. The average age of the injured worker was 35.2 years. Costs per injury increased as age increased. The average return to work period for lost time claims was 48.9 days. The highest proportions of injuries were sprains and strains (30.4%) and the most common location of injury were extremities (63.5%).

A large proportion of low cost, low severity injuries were observed in the dataset. More importantly, a decreasing trend in total claims and injury rate were also observed. A major concern, however, is the total cost of paying claims continues to rise.

When comparing these results to national data, injury rates are comparable. Goldcamp (2010) found similar demographic results and injury rates for farm employees in 2001 and 2004. Myers (2001) observed similar injury rates and leading types of injuries in farm operators.

Ohio BWC data can serve multiple roles. At the fundamental level, BWC data performs actuarial functions. However, it also has utility to safety and health professionals

Recommendations

Further work can be done to improve workers' compensation data for describing the status of agricultural injury: 1) Improvements to data collection and coding to increase the rate of accurate claim capture and usefulness, 2) Universal coding for industry or risk groups, 3) A systematic reporting system to ensure active surveillance of injuries as they occur, instead of retrospectively querying data, and 4) Coupling workers' compensation data with records from state coroners, EMS and hospital records, and the Occupational Safety and Health Administration, as well as other stakeholders.

Conclusions

Ohio BWC data can be used to determine costs associated with injury related to agriculture. Injury rates for workers covered under Ohio BWC insurance can be calculated using the methods identified in this paper. Finally, Ohio BWC data can provide accurate and precise conclusions to help safety and health professionals guide intervention programs.

References

Barell, V., Aharonson-Daniel, L., Fingerhut, L. a, Mackenzie, E. J., Ziv, A., Boyko, V., Abargel, A., et al. (2002). An introduction to the Barell body region by nature of injury diagnosis matrix. *Injury prevention : journal of the International Society for Child and Adolescent Injury Prevention*, 8(2), 91-6. Retrieved from http://www.pubmedcentral.nih.gov/articlerender.fcgi?artid=1730858&tool=pmcentrez&rendertype=abstract

Demers, P., & Rosenstock, L. (1991). Occupational injuries and illnesses among Washington State agricultural workers. *American journal of public health*, 81(12), 1656-8. Retrieved from http://www.pubmedcentral.nih.gov/articlerender.fcgi?artid=1405289&tool=pmcentrez&rendertype=abstract

Fleiss, J. L., Levin, B., & Paik, M. C. (2004). The Measurement of Interrater Agreement. Statistical Methods for Rates and Proportions (Third., pp. 598-626). Hoboken, NJ: John Wiley & Sons, Inc. doi:10.1002/0471445428.ch18

Goldcamp, E. M. (2010). Work-Related Non-Fatal Injuries to Adults on Farms in the U.S., 2001 and 2004. *Journal of Agricultural Safety and Health*, 16(December 2009), 41-51.

Myers, J. R. (2001). Injuries Among Farm Workers in the United States - 1995. *Safety And Health* (p. 338). Cincinnati, OH. Retrieved from http://www.cdc.gov/niosh/docs/2001-153/pdfs/2001-153.pdf

Ohio Bureau of Workers' Compensation, (2011). Ohio BWC State Insurance Fund Manual (July 1, 2011 - June 30, 2012). Spring (p. 273). Columbus, OH. Retrieved from http://www.ohiobwc.com/downloads/blank-pdf/StateFundManual.pdf

Ohio Department of Agriculture, . (2010). 2010 Annual Report of the Ohio Department of Agriculture. *The American psychologist* (Vol. 66, pp. S1-48). Columbus. doi:10.1037/a0024196

U.S. Department of Labor, . (2010). 2010 Survey of Occupational Injury & Illnesses Summary Estimates Charts Package. October, 16.

United States Department of Agriculture - National Agricultural Statistics Service, . (2009). 2007 Census of Agriculture - Ohio State and County Data. Area (Vol. 1, p. 747). Retrieved from http://www.agcensus.usda.gov/Publications/2007/Full_Report/Volume_1,_Chapter_1_State_Level/Ohio/ohv1.pdf

Villarejo, D. (1998). Occupational Injury Rates Among Hired Farmworkers. *Journal of Agricultural Safety and Health*, 4(5), 39-46. Retrieved from http://elibrary.asabe.org/azdez.asp?JID=3&AID=15373&ConfID=j1998&v=4&i=5&T=2&redirType=

Table 1. Occupational Groups Queried to Yield Initial Data

Occupational Group Description
Farm: Nursery Employees & Drivers
Farm: Gardening – Market or Truck - & Drivers
Farm: Orchard or Grove & Drivers
Farm: Poultry or Egg Producer & Drivers
Farm: Florist & Drivers
Farm: Dairy & Drivers
Farm: Field Crops & Drivers
Farm: Berry or Vineyard & Drivers
Farm: Cattle or Livestock Raising NOC & Drivers
Farm: Fish Hatchery & Drivers
Farm: Animal Raising & Drivers
Irrigation Works Operation & Drivers
Logging or Tree Removal – Non-mechanized Operations
Logging or Tree Removal –Mechanized Operations
Stable or Breeding Farm & Drivers

Table 2. Aged 30 Month Medical Costs per Manual (2000-2008)

Description	Total Cost	No. Claims	Cost/Claim
Logging or Tree Removal - Mechanized Operations	$ 136,271.67	5	$ 27,254.33
Logging or Tree Removal - Non-Mechanized Operations	$ 12,539,921.50	544	$ 23,051.33
Farm: Field Crops & Drivers	$ 17,562,537.06	1526	$ 11,508.87
Farm: Dairy & Drivers	$ 9,324,659.91	980	$ 9,514.96
Stable or Breeding Farm & Drivers	$ 8,253,882.86	945	$ 8,734.27
Farm: Cattle or Livestock Raising NOC & Drivers	$ 7,871,035.56	1117	$ 7,046.59
Farm: Orchard or Grove & Drivers	$ 890,558.42	130	$ 6,850.45
Farm: Berry or Vineyard & Drivers	$ 363,204.74	58	$ 6,262.15
Farm: Florist & Drivers	$ 6,826,673.71	1564	$ 4,364.88
Farm: Poultr or Egg Producer & Drivers	$ 6,489,077.54	1597	$ 4,063.29
Farm: Gardening - Market or Truck & Drivers	$ 2,688,597.07	717	$ 3,749.79
Farm: Animal Raising & Drivers	$ 70,586.38	23	$ 3,068.97
Farm: Nursery Employees & Drivers	$ 7,365,478.44	2919	$ 2,523.29
Farm: Fish Hatchery & Drivers	$ 8,493.19	12	$ 707.77
Irrigation Works Operation & Drivers	$ 249.34	1	$ 249.34
Total	**$ 80,391,227.39**	**12138**	**$ 6,623.10**

Figure 1. Flow Chart of Claim Numbers Used for Calculations

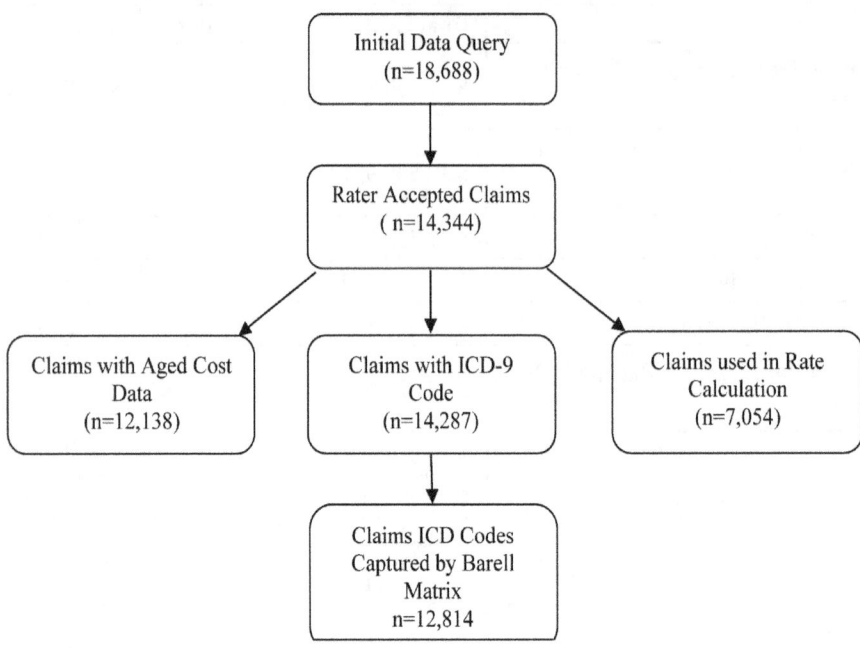

Figure 2. Agricultural Injuries and Rates (per100 FTE)

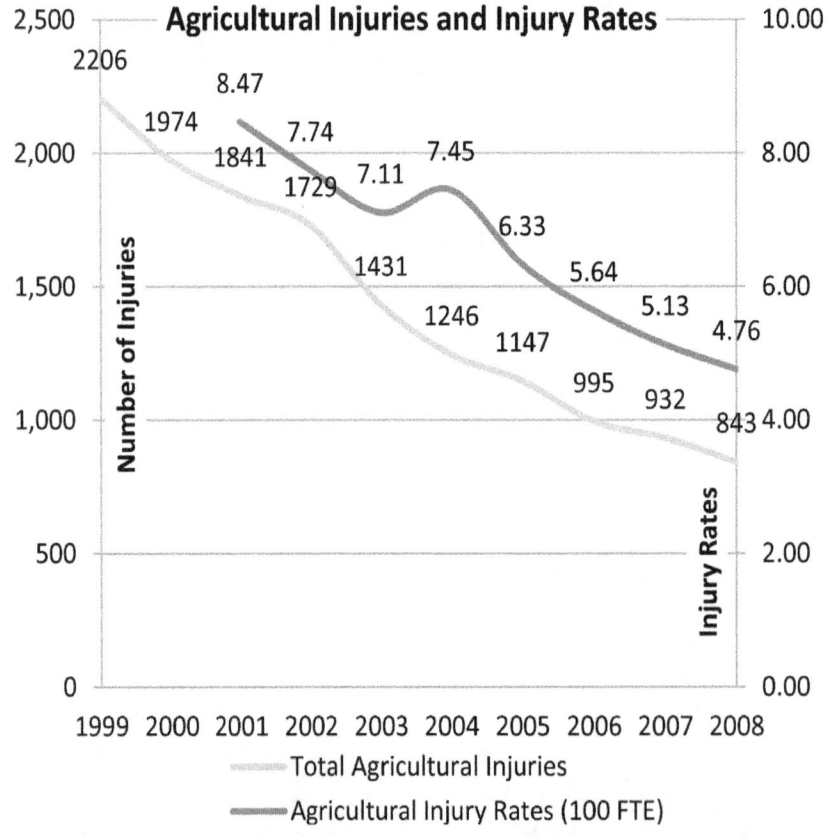

Use of Multiple Data Sources to Enumerate Work-Related Amputations in Massachusetts: The Contribution of Workers' Compensation Records

L Davis, K Grattan, S. Tak, L Bullock, L. Boden
Massachusetts Department of Public Health

ABSTRACT

Background
Accurate surveillance information about work-related (WR) injuries is essential to guide prevention efforts. Innovative approaches that combine information from multiple sources are needed. Massachusetts piloted multi-data source surveillance of WR amputations.

Methods
WR amputations during 2007-2008 were identified using the Massachusetts Survey of Occupational Injuries and Illnesses (SOII) sample and four administrative data sets: workers' compensation indemnity claim records (WC) and inpatient, outpatient, and emergency department data –collectively referred to as Case Mix (CM) data. Potential WR amputations were identified using injury and procedure codes and narrative text searches. Payment by workers' compensation was used to identify potential work-related cases in the CM dataset. Medical records for potential CM cases were also obtained abstracted to obtain personal/employer identifiers and assess work-relatedness. Data sets were linked and case inclusion criteria were applied to enumerate a final statewide count of cases. Capture rates by data source and characteristics of cases captured by different sources were examined.

Results (Preliminary)
Application of selection criteria yielded 447 and 2,630 potential WR amputations in WC and CM data, respectively, and approximately 60 cases in the SOII sample. After matching and application of inclusion criteria, approximately 800 WR amputations were identified. Approximately 42% of the cases were captured by WC records; 14% were identified by WC records only. Payment by WC was a highly predictive of work-relatedness in CM but failed to capture approximately 20% of work-related cases treated in hospitals. The total count of approximately 800 far exceeded the 210 cases estimated by SOII for 2007-08. Some cases were reported in the SOII as other injuries and some were not eligible for inclusion. Reasons for ineligibility will be presented.

Conclusions
Use of multiple data sources identified substantially more WR amputations than any single source. Information from the state workers' compensation system contributed substantially to the overall count. This included information from workers' compensation records and information about payment by workers' compensation in administrative records maintained by hospitals. Multi-source surveillance enhances our ability to characterize injury burden but also poses practical challenges that will vary by state.

Workers' Compensation-Related CSTE Occupational Health Indicators

Erin Simms§, Caroline Tai§, Meredith Towle* and Kenneth Rosenman#
§Council of State and Territorial Epidemiologists, * Colorado Department of Public Health and Environment, #Michigan State University

Introduction

Workers' Compensation (WC) data is a resource that is widely available to state and local health departments. Carpal tunnel syndrome and amputations are two often well-documented conditions where claims with days away from work are recorded. Both carpal tunnel syndrome and work-related amputations are preventable, and the control of occupational hazards within the workplace is the most effective means of prevention. Estimating the burden of such occupational illnesses and injuries and tracking these occurrences over time aid in targeting public health prevention programs and activities, including new regulations. Information on reported cases of work-related injury or illness can be used to identify contributory factors and to develop improved or new prevention strategies or regulations to protect workers.

Accepted WC awards represent known work-related injuries and illnesses, and potentially the more severe cases. The total and average amounts of WC benefits paid in a state or region estimate the financial burden of these events, which further illustrates the importance of prevention programs and activities.

There is, however, variability among state WC coding systems for eligibility, completeness, and quality control. For example, some state workers' compensation agencies collect only the subset of 'claims' legally contested, while others do not. WC data cannot be relied on for a complete picture of work-related injury or illness, as many individuals with work-related illnesses and injuries do not file for workers' compensation (1,2). Additionally, WC claims may be accepted or denied, and state rules for classifying claims as lost time vary. Finally, self-employed individuals such as farmers, small business owners, independent contractors and federal employees, railroad, longshore and maritime workers are not covered by state workers' compensation systems.

The Council of State and Territorial Epidemiologists (CSTE) Occupational Health Indicators address the need for improved consistency and availability of occupational disease and injury surveillance data.

Methods

In 1998, CSTE and the National Institute for Occupational Safety and Health (NIOSH) convened a work group to make recommendations to NIOSH concerning state-based surveillance activities for the coming decade. The Work Group identified the need for improved consistency and availability of occupational disease and injury surveillance data and developed a standard set of 20 "Occupational Health Indicators" (OHIs), which are tracked currently by 28 states annually.

There are 3 CSTE OHIs that utilize data from WC claims:
- Indicator 5 – State WC claims for amputation with lost work-time
- Indicator 8 – State WC claims for carpal tunnel syndrome with lost work-time
- Indicator 19 – WC awards

Data for Indicators 5 and 8 are obtained from individual state worker compensation agencies.

---

[1] A lost time claim is one in which the worker misses a certain number of calendar days, work days or work shifts due to the injury or illness.

The data source for Indicator 19 is the National Academy of Social Insurance (NASI). A description and a how-to-guide for calculating each indicator is in the CSTE document titled Occupational Health Indicators: A Guide for Tracking Work-Related Health Conditions and Their Determinants, and available on the CSTE website at www.cste.org (3).

Results

Due to the variances in state WC claims coding systems, we are unable to draw conclusions from state-to-state comparisons.

Indicator 5 - State workers' compensation claims for amputation with lost work-time
The annual incidence rate of amputation cases is calculated using cases of amputations with lost work-time filed with the state workers' compensation system as the numerator divided by the total number of workers covered by the compensation system. Figure 1 illustrates the most recent data for Indicator 5 by state for 2008.

Indicator 8 – State workers' compensation claims for carpal tunnel syndrome with lost work-time
The annual incidence rate of carpal tunnel syndrome cases is calculated using cases of carpal tunnel syndrome with lost work-time filed with the state workers' compensation system as the numerator divided by the total number of workers covered by the compensation system. Figure 2 illustrates the most recent data for Indicator 8 by state for 2008.

Indicator 19 – Workers' compensation awards
The average annual amount of workers' compensation benefits paid per covered worker is calculated by taking the amount of workers' compensation benefits paid by state and dividing it by the total number of civilians employed aged 16 years and older. Figure 3 illustrates the most recent data for Indicator 19 by state for 2008.

State Perspective: The Colorado Department of Public Health and Environment (CDPHE) is able to obtain aggregate WC data for reporting Indicators 5 and 8. Figures 4 and 5 illustrate these data for 2001-2009. These data represent only accepted claims. The waiting period for lost-time status is defined as >3 calendar days or >3 shifts way from work (Colorado Revised Statute 8-43-101)

Workers who are self-employed and federal employees are not covered by Colorado workers' compensation insurers and therefore are not included in these estimates. However, the NASI covered worker data used for rate calculations do include government workers. Also, there may be a lag time in reporting claims. In Colorado, an average of 80% of claims are filed in the year the injury or illness occurs.

Recommendations from Colorado's perspective include: continue to utilize WC and other data sources to monitor trends in work-related injury and illness; further analyze and describe existing OHI WC claims data by occupation, industry, age, gender, and cost; promote opportunities to add race/ethnicity variables to WC claims data; and improve completion rate of industry and occupation data captured in WC claims.

Discussion

There are several benefits to using WC data for occupational health surveillance:
- Rich data source for state-based occupational health surveillance
- Relative ease with which such administrative data and the costs associated with these claims can be accessed
- Data sharing may increase collaboration among state-agencies (i.e. between departments of health and labor/employment)

There are also many challenges, which currently limit states use of WC data:
- Differences in the availability of data (i.e., for lost time cases only versus all medical benefits cases), benefit eligibility criteria, and compensation scales between states indicate that claims and cost data should be used to evaluate trends within a state and

can only be used to make state-to-state comparisons if this variability is accounted for
- Lack of data on race and ethnicity
- Data on industry/occupation may be incomplete or not coded for analysis
- WC data my not distinguish between claims accepted and denied
- Restrictions may be placed on data use (i.e. Linking data to other health data sets or using data to target OSHA enforcement)
- Likely undercount because injured workers may not apply or are not eligible to apply (i.e., self-employed)
- ICD codes not used as in medical data bases

References

1. Bonauto D, Largo T, Rosenman K, Green M, Walter J, Materna B, Flattery J, St. Louis T, Yu L, Fang S, Davis L, Valiante D, Cummings K, Hellsten J, Prosperie S [2010] Proportion of Workers Who Were Work-Injured and Payment by Workers' Compensation Systems - 10 States, 2007, MMWR 59(29), pp 897-900.

2. Rosenman KD, Gardiner JC, Wang J, Biddle J, Hogan A, Reilly MJ, Roberts K, Welch E [2000] Why Most Workers with Occupational Repetitive Trauma Do Not File for Workers' Compensation, J Occup Environ Med 42(1):25-34.

3. The Council of State and Territorial Epidemiologists [2012] Occupational Health Indicators: A Guide for Tracking Occupational Health Conditions and Their Determinants. http://www.cste.org/webpdfs/Occupational/OHIndicatordocumentJune2012.pdf.

Figure 1. Rate of Lost Work Time Claims for Amputations Identified in Workers' Compensation Systems by State, 2008

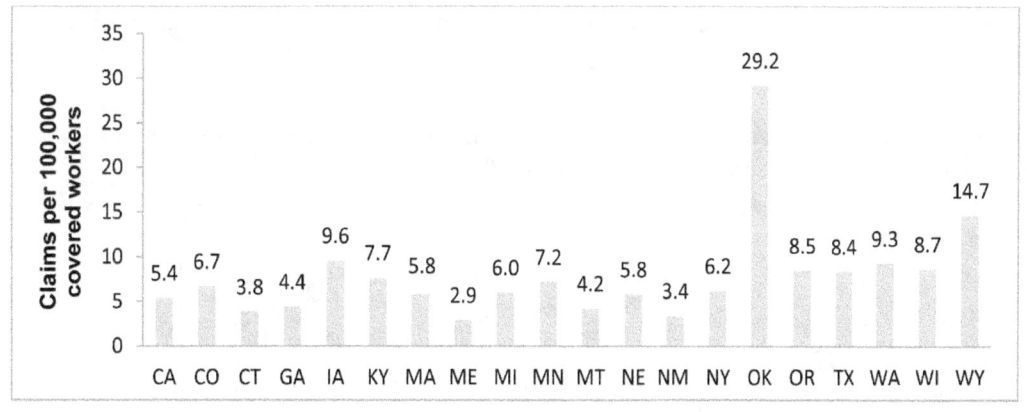

Figure 2. Rate of Lost Work Time Claims for Carpal Tunnel Syndrome Cases Identified in State Workers' Compensation Systems by State, 2008

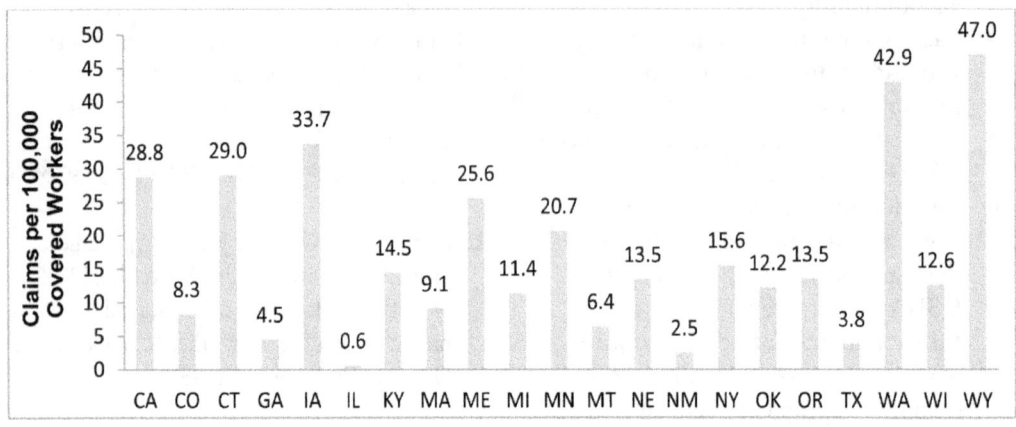

Figure 3. Average Workers' Compensation Benefit Paid per Covered Worker* by State and U.S., 2008

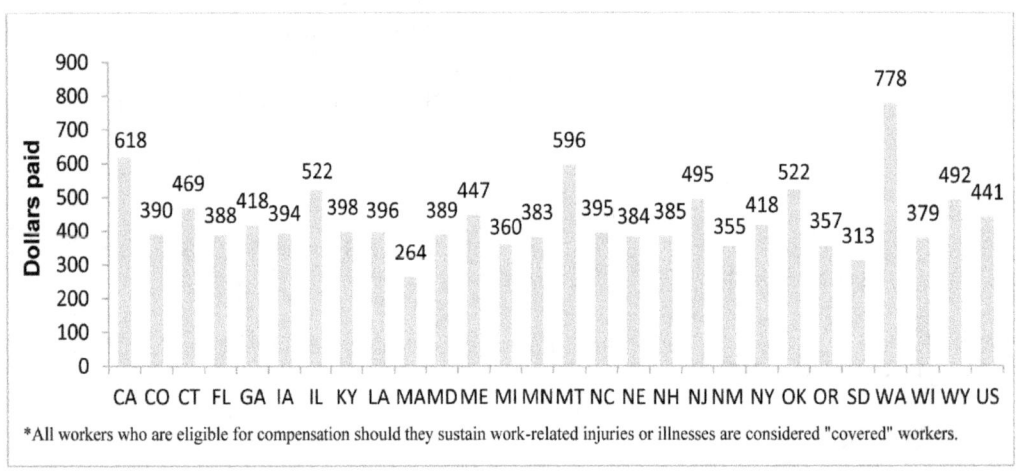

*All workers who are eligible for compensation should they sustain work-related injuries or illnesses are considered "covered" workers.

Figure 4. Annual incidence rate of amputation claims filed with state workers' compensation per 100,000 workers covered, Colorado, 2001-2009*

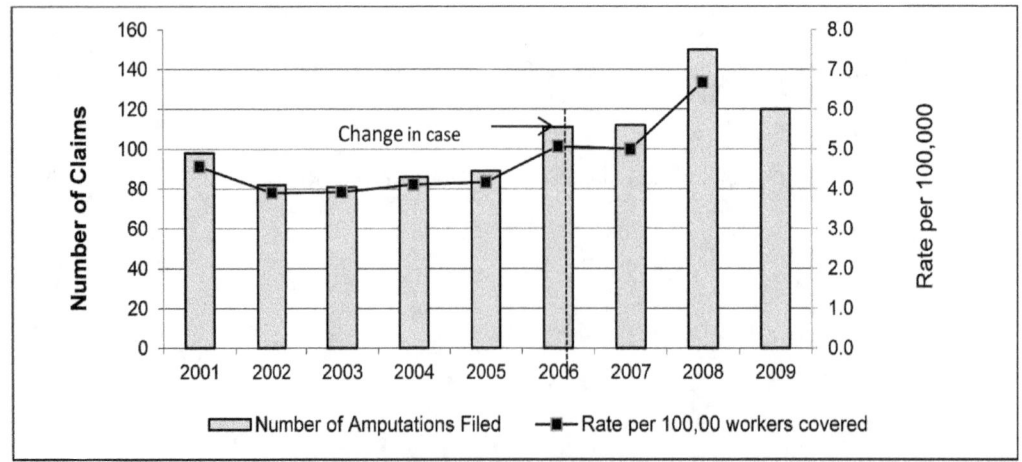

Figure 5. Annual carpal tunnel syndrome cases filed with state workers' compensation per 100,000 workers covered, Colorado, 2001-2009*

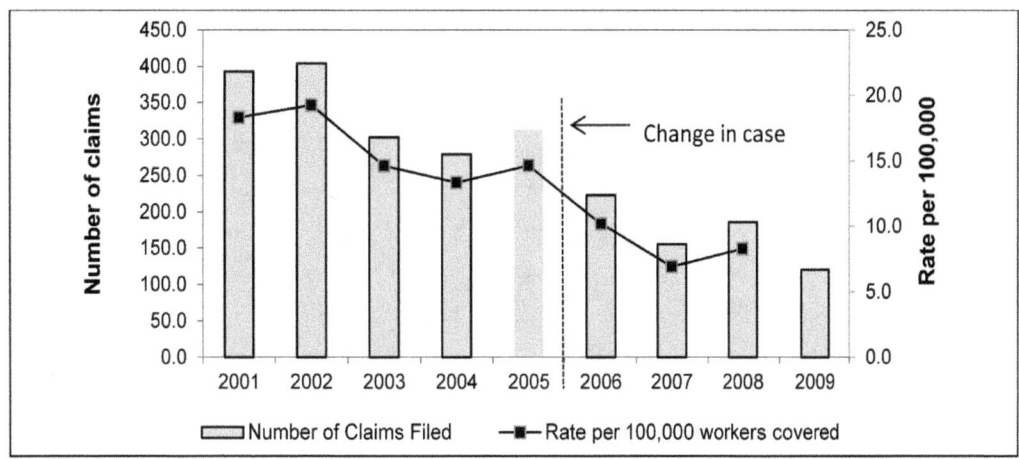

The Effectiveness of the Safety and Health Achievement Recognition Program (SHARP) in Reducing the Frequency and Cost of Workers' Compensation Claims

Ibraheem Tarawneh, Ph.D., Donald Bentley, PE, CIH, Michael Lampl, MS, CPE, David Robins
Ohio Bureau of Workers' Compensation Division of Safety and Hygiene

Background

The majority of research exploring the effectiveness of safety and health management systems (SHMS) in reducing injury rates reported positive, yet varied outcomes (Robson et al., 2005). Research in this area is challenging due to the wide scope of these systems, the many variables involved in implementing them, and the variety of factors that can affect injury rates. However, generally, the literature indicates that these systems, depending on the system elements and level of implementation, can reduce occupational injury rates (Arocena, et al. 2010; Robson et al., 2005; 2007). The objective of this study was to explore the effect of implementing SHMSs on workers' compensation (WC) frequency and cost of claims by studying the claim experience of companies that achieved the Safety and Health Achievement Recognition Program (SHARP) status as prescribed by the Occupational Safety and Health Administration (OSHA). SHARP distinguishes small employers implementing and operating exemplary SHMSs. Companies can only achieve SHARP status through working closely with the Occupational Safety and Health Administration On-site Consultation Program (OCP). OCP is primarily funded by Federal OSHA in collaboration with participating states to provide occupational safety and health consultative services to small employers (less than 250 employees) in high hazard industries. In 2005, the Ohio OCP became part of the Division of Safety and Hygiene; the loss prevention arm of the Ohio Bureau of Workers Compensation (BWC). This provided a great opportunity for studying the claim experience for employers who achieved SHARP recognition between 2004 and 2010.

Methods

The study included a review and analysis of the Safety & Health Program Assessment Worksheet (OSHA Form 33) scores, as well as claims and injury experience including frequency, medical and indemnity costs along with reserves (30-month incurred cost) for 16 Ohio companies that achieved SHARP recognition between the years 2004 and 2010.

Description of Companies Included in the Study

Table 1 provides basic information about the companies included in the study including the type of business based on their primary North American Industry Classification System (NAICS) code, the average number of employees during a year, and the year in which they entered SHARP.

OSHA

Form 33 Scores

In order to achieve SHARP status a comprehensive consultation is performed by OCP consultants utilizing the OSHA Form 33. Form 33 consists of three major components, seven subcomponents, and 58 attributes of an organizational SHMS. Attributes are measures of safety and health implementation that together form a comprehensive SHMS. For

---

[1] The findings and conclusions in this paper are those of the authors and do not necessarily represent the views of the Ohio Bureau of Workers' Compensation

example under the subcomponent Hazard Anticipation and Detection one attribute is: An effective hazard reporting system exists. This breakdown of attributes is shown in Table 2. OCP consultants would rate an employer for each attribute on a scale of "0" to "3." The scoring legend for the attributes is as follows (0 = No; 1 = No, Need major improvement; 2 = Yes, Needs minor improvement; and 3 = Yes). The before and after achieving SHARP Form 33 data was assembled and analyzed for the 16 companies according to each attribute and subcomponent. It is worth noting that to achieve SHARP, in addition to receiving satisfactory scores for the attributes on Form 33, the employer must maintain Days Away, Restricted, or Transferred (DART) rate and Total Recordable Case (TRC) rate below the national average for their respective industry.

Frequency and Cost of Claims
The data relative to frequency and cost of claims was assembled for the 16 companies over a period of six years. The period of six years included the experience of each company in the 24 months prior to working with OCP to achieve SHARP, the 18 to 24 months while working with OCP consultants, and the 24 months after achieving SHARP. The study years were limited to the three segments of 24 months each, because a significant number of the companies achieved SHARP in 2008, 2009, and 2010; along with the fact that it took most of the companies between eighteen and twenty-four months to achieve SHARP recognition. Actual 30 month incurred costs were not available for 2010 and 2011 claims, accordingly a predicted cost was used for claims occurring in those years based on the trend of previous claim years. Because of a change to the BWC reserving system in 2008 that resulted in lowering claims reserves at the global system level, for the purposes of this study, the cost of claims was increased by 25% for claims that occurred after the change.

Results
OSHA Form 33 Scores
The before and after achieving SHARP average scores for each of the seven sub-components of Form-33 for the 16 companies included in the study are shown in Figure 1. Figure 1 clearly shows the lack of uniformity in the average scores among the seven sub-components before achieving SHARP. Generally, before achieving SHARP, these companies achieved relatively higher scores on attributes related to hazard prevention and control, administration and supervision, and management leadership. On the other hand, the companies had relatively lower scores on attributes related to planning and evaluation, safety and health training, and employee participation.

After achieving SHARP, the companies improved their scores in terms of higher scores and better uniformity among all seven sub-components. The average scores among the seven subcomponents were about 2.5 out of 3 after achieving SHARP.

Frequency and Cost of Claims
The results have shown considerable decrease in frequency and cost of claims from before to after achieving SHARP recognition. For the sixteen companies, the total number of claims was 213 for the two year period prior, 229 for the two year period while working toward SHARP, and 128 for the two year period after achieving SHARP. This represents a 40% decrease in claims for the two year period after the companies achieved SHARP. The claims frequency results are shown in Table 3.

The total 30-month incurred cost of claims went down 84% from $2,124,387 before SHARP to $336,047 after SHARP. The 30-month incurred cost per one million dollar of payroll went down from $17,041 to $2,461, which represents about 86% decrease. The cost of claims results are shown in Figure 2.

Conclusion
Relative to Form-33 scores, many of the companies achieved relatively acceptable scores in the majority of the sub-components. However, working toward and achieving SHARP resulted in not only improving the companies' scores in each of the seven sub-components, but also in achieving uniform

performance among all subcomponents. Achieving SHARP recognition resulted in significant reductions in the frequency and cost of claims. Although the total number of claims slightly increased during the two-year period while the companies were working toward SHARP, the total cost of claims during the same period decreased by almost 58%. This may indicate better reporting of accidents and injuries with the majority of injuries characterized by lower severity. On the other hand, in the two years after achieving SHARP, the frequency decreased by about 40% from that in the two years before achieving SHARP and the total cost decreased by almost 84% over the same period of time. The preliminary results from this study show a significant value for achieving SHARP recognition for small employers in high hazard industries and its effect on reducing the frequency and cost of workers' compensation claims. Analysis of Form 33 data show that, before achieving SHARP, companies had considerable deficiencies in employee participation, planning and evaluation and safety and health training subcomponents. The value in an exceptional SHMS is realized in reductions in both injury frequency and cost. Although accounted for based on global measures within the BWC system, the study results are somewhat limited due to the changes in the reserving system as well as the fact that seven of the sixteen companies examined achieved SHARP in 2009 and 2010.

References

Arocena, P.; Nunez, I. 2010. An empirical analysis of the effectiveness of occupational health and safety management systems SMEs. *International Small Business Journal* 28(4): 398-419.

Robson, L.; Clarke, J.; Cullens, K.; Bielecky, A.; Severin, C.; Bigelow, P.; Irvin, E.; Culyer, A.; Mahood, Q. 2005. The effectiveness of occupational health and safety management systems: A systematic review. A technical report. Institute for Work and Health, Toronto.

Robson, L., Clarke, J., Cullens, K., Bielecky, A., Severin, C., Bigelow, P., Irvin, E., Culyer, A., Mahood, Q., 2007. The effectiveness of occupational health and safety management systems: a systematic review. Safety Science 45: 329–353.

Robson, L.; Speers, J.; Kusiak, R.; Burns, B. 2007. Development of a performance measurement report for the Ontario prevention system. *Policy and Practice in Health and Safety* 05 (1).

Table 1. Basic Information about the companies examined.

Company	Type of Business	Number of Employees	Year Achieving SHARP
A	Spring (Light Gauge) Manufacturing	148	2008
B	Metal Service Centers & Other Metal Merchant Wholesalers	39	2008
C	Hazardous Waste Treatment and Disposal	137	2009
D	General Warehousing and Storage	63	2010
E	Miscellaneous Wood Product Manufacturing	29	2008
F	Folding Paperboard Box Manufacturing	124	2010
G	Iron, Steel Pipe, & Tube Manufacturing from Purchased Steel	40	2010
H	Metal Service Centers & Other Metal Merchant Wholesalers	86	2009
I	Research and Develop in the Physical, Eng., & Life Sciences	34	2007
J	Steel Wire Drawing	22	2009
K	Other Rubber Product Manufacturing	36	2005
L	Commercial and Service Industry Machinery Manufacturing	42	2004
M	Other Plastics Product Manufacturing	23	2008
N	Other Warehousing and Storage	28	2006
O	Postharvest Crop Activities (except Cotton Ginning)	50	2010
P	Blankbook, Looseleaf Binders, and Devices Manufacturing	121	2007

Table 2. OSHA Form 33 subcomponents.

Component	Subcomponent	Attributes
Operational	Hazard Anticipation and Detection	1-10
Operational	Hazard Prevention and Control	11-19
Managerial	Planning and Evaluation	20-25
Managerial	Administration and Supervision	26-33
Managerial	Safety and Health Training	34-39
Cultural	Managerial Leadership	40-49
Cultural	Employee Participation	50-58

Table 3. Frequency of claims over six years.

Company	Before Year 1	Before Year 2	Working Toward SHARP Year 1	Working Toward SHARP Year 2	After Year 1	After Year 2	Total
A	32	22	30	26	21	6	137
B	19	12	18	17	11	14	91
C	12	12	9	12	5	8	58
D	8	8	6	8	2	2	34
E	10	11	10	12	7	5	55
F	8	6	6	8	5	6	39
G	2	6	2	3	3	1	17
H	9	6	4	3	3	2	27
I	0	1	3	6	2	4	16
J	1	2	8	4	4	3	22
K	0	0	3	2	0	4	9
L	4	3	2	0	0	1	10
M	2	0	2	2	1	1	8
N	0	1	1	1	0	0	3
O	0	0	2	0	0	1	3
P	11	5	10	9	4	2	41
Total	**118**	**95**	**116**	**113**	**68**	**60**	**570**

Figure 1. Average scores for all companies included in the study by subcomponent.

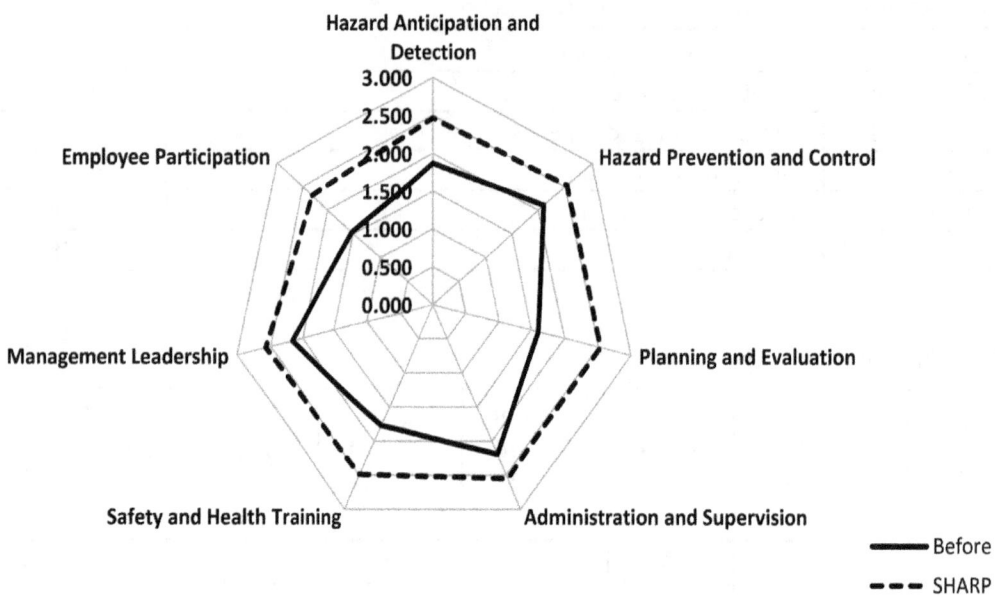

Figure 2. Claims costs and costs per $1.0 million payroll over six years.

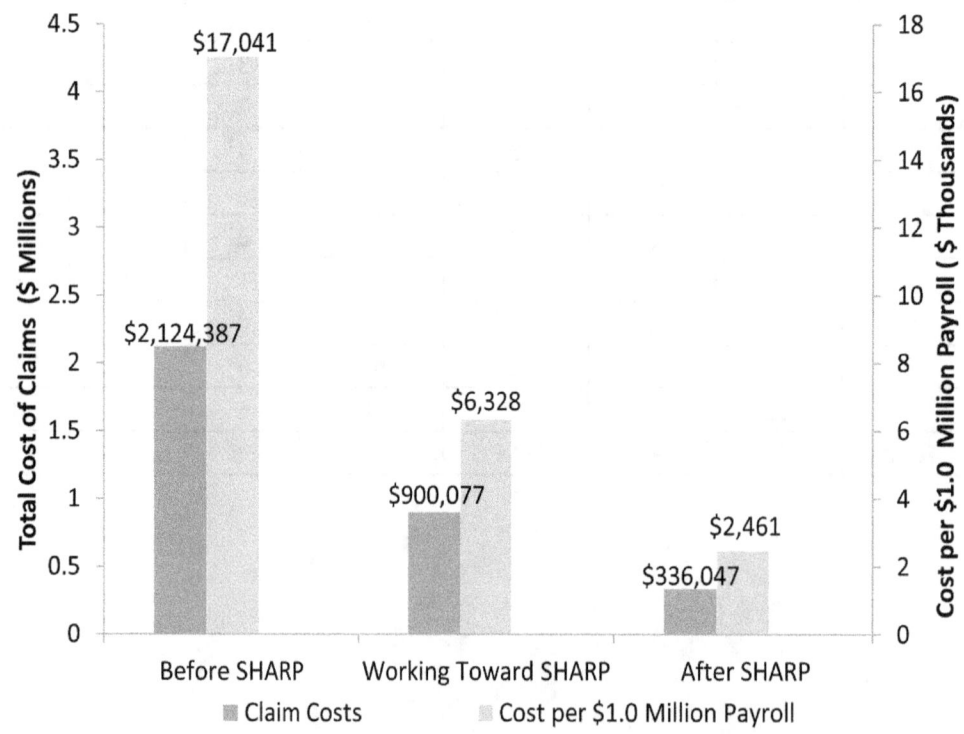

Comparison of Cost Valuation Methods for Workers Compensation Data[1]

Wurzelbacher SJ§, Meyers AR§, Bertke SJ§, Lampl M*, Robins DR*, Bushnell TP§, Tarawneh A*, Childress D*, Turnes J*

§National Institute for Occupational Safety and Health, *Ohio Bureau of Workers Compensation

Background

The purpose of this study was to compare and contrast three workers compensation (WC) claim cost-valuation methods using data from the Ohio Bureau of Workers Compensation (OBWC). WC claims are composed of several cost components including payments for medical procedures, payments for indemnity (replacement wages), and claim reserves, which are anticipated future medical and indemnity costs. All paid and reserve costs combined for the claim are defined as the "total incurred" claim cost. WC claims can be open for extended periods of time, such that payments can be made over the course of months or years. As a claim ages, reserve cost totals diminish as the reserves are converted to paid totals. Once a claim is closed, the reserve cost is $0.

The "Most-Recent" method calculates costs as of a recent available date. This method is often used by insurers for both individual and aggregated claims to report losses back to insured clients. The purpose is to track current expected costs for the company and benchmark to insured peers for a given time period based on industry type and company size. Although the Most-Recent method is one of most commonly applied methods for benchmarking purposes, a possible basic drawback with the method for evaluating cost trends over time is that older claims are allowed more time to develop costs than newer claims. This means that a higher proportion of the total cost of older claims will be actual costs paid to date rather than reserves for future costs. This will bias the total cost estimate of older claims relative to more recent claims. At OBWC, reserve amounts for each claim are calculated using the proprietary "MIRA" system. These reserves represent estimates of the most likely future cost of the claim, which approximates the mode of the distribution of claims of that type, rather than the mean (or expected value). The Most-Recent method does not include inflation adjustments.

The "30-Month" method calculates costs after claims have been aged for a more consistent period of time. Costs of all claims are valued 30 months after January 1 of the calendar year in which the claim occurred (i.e. each claim is aged between 18-30 months). This method is used for both individual and aggregated claims to represent cost trends over time and to evaluate the effectiveness of interventions (e.g. compare losses before and after implementation). The 30-Month method is specifically designed to address the issue of differences in the valuation of claims associated with differences in the age of claims. The 30-Month method, like the Most-Recent method, includes reserve amounts that are estimates of the mode (most likely) future cost of claims of the same type, and does not include inflation adjustments.

One potential drawback of the 30-Month method is that the claim values are locked into past values that may be reflective of insurer system characteristics that were operating at that time. For example, OBWC changed reserving systems (e.g. MIRA I to MIRA II) such that, for the 30-Month method, reserves for claims prior to 2007 were calculated using

---

[1] The findings and conclusions in this paper are those of the authors and do not necessarily represent the views of the Ohio Bureau of Workers' Compensation or the National Institute for Occupational Safety and Health.

MIRA I and reserves for claims 2007 and after were calculated using MIRA II. In contrast, the Most-Recent method (if applied 2007 and after) uses MIRA II for all claims, even for claims prior to 2007. OBWC has determined that MIRA II generally calculates smaller reserves for the same type of claim compared to MIRA I. Therefore, if trends over time using 30-Month reserve costs span the 2007 period, trend estimates will be biased downward, reflecting reserve system changes as well as changes of interest (e.g. industry exposure changes or intervention effects etc.).

A drawback of both the 30-Month and Most-Recent methods is that they do not represent the best estimate of the absolute values of claims. The Factor-Adjusted method addresses this limitation. It calculates costs by applying actuarial loss development factors that attempt to estimate the ultimate payout amounts for the claims. Reserves therefore represent the mean future cost of claims of the same type. This method is used by insurance underwriters for the purpose of analyzing aggregated claims for loss trends. A potential drawback with the Factor-Adjusted method is that it is intended to be applied to groups of claims, and its values are usually higher than actual individual claim values (since the mean is always higher than the mode and median in claim cost distributions). Another limitation of the Factor-Adjusted method is that the factors being used are based on all OBWC claims, so applying the factors to one industry may distort results. Unlike the other methods, the Factor-Adjusted method includes inflation adjustments for medical payments and projections of future costs are stated not in current year dollars, but dollars of future years.

Methods

Cost data from over 61,000 claims valued using the three methods were downloaded for all single location, OBWC-insured wholesale/retail trade (NAICS 42, 44, 45) companies for calendar years 2004-2009. Cost data included values for paid medical treatments, paid indemnity (compensation payments for lost wages), and reserved costs for the claim. The valuation date for the Most-Recent method was 12/31/2011, and slightly earlier, as of 9/30/2011, for the Factor-Adjusted method. OBWC sponsors two main programs that impact the cost of claims reported in their database. The first program allows insured companies to pay first dollar medical costs up to a specified limit for medical-only claims. Only medical paid costs in excess of this limit are reported to OBWC. A second program allows insured companies to pay first dollar indemnity costs, which are not reported to OBWC. Claims that were affected by either of these two programs were excluded from the cost comparison analyses. Many of the claims affected by these programs had a reported cost of zero, but there were also claims not affected by these programs that also had zero cost. These were also excluded from the analysis. To be defined as a $0 claim, the claim had to have a value of $0 for all three valuation methods. The three methods were first compared by calculating and comparing total incurred claim costs (medical paid + indemnity paid + reserves) as estimated by each method in each year 2004-2009, and for the 2004-2009 period as a whole. To test the statistical significance of the cost differences between methods, the non-parametric Wilcoxon Signed Rank test was used to compare the differences on a claim-level basis.

Next, 2004-2009 cost trends based on cost estimates of the three methods were compared. Two types of cost trends were calculated: trends in total cost per claim and trends in total cost per employee. The unit of analysis for cost per claim trends was the individual claim. The unit of analysis for cost per employee trends was the WRT industry subsector (3-digit NAICS code). Year-over-year rate ratios were calculated, with 95% confidence intervals. A single rate ratio for each method was calculated, but the average cost per claim or employee was allowed to vary by 3-digit NAICS code, since subsectors differ widely in costs.

Cost per claim was modeled using a log transformation, since the distribution of individual claim costs was highly skewed, with a large proportion of low cost claims. The log transformation reduces the impact of very high cost, outlier claims which, without the log transformation, were observed to have large effects upon trend estimates and create an amount of year-to-year variation that makes trends more difficult to detect. Results of the trend analysis of logged claim cost are most accurately expressed in terms of the geometric mean, which usually varies in a way similar to the median in distributions with a strong rightward skew. Cost per employee was modeled as a rate, using negative binomial regression, which is robust to the distributional form of the cost. The results are presented in terms of mean cost per employee. All analyses were conducted using SAS version 9.2 (SAS Institute, Inc., Cary, NC).

Results

Costs per claim: The methods produced significantly different ($p < 0.0001$) total incurred values compared to each other on a claims level basis. Table 1 provides a summary of mean and median claim values with each method.

Geometric mean cost per claim trends: The three valuation methods yielded trends in total incurred cost that were substantially different, although the only differences that were statistically significant were those between the Factor-Adjusted cost trend and the trends using the other methods (Figure 1). Geometric mean costs increased by 24.9%, 15.8%, and 29.3% for the 30-Month, Most-Recent, and the Factor-Adjusted methods, respectively, from 2004-2009.

Mean cost per employee trends: While trends in total incurred costs were somewhat different, none of these differences were statistically significant (Figure 2). The mean cost per employee decreased by 46.2%, 56.3%, and 34.2% for the 30-Month, Most-Recent, and Factor-Adjusted methods respectively from 2004-2009.

Discussion

This study indicated that the three valuation methods tested produced different total incurred claim costs, total incurred cost per claim trends, and total incurred cost per employee trends. The importance of these findings depends upon the intended use of the data.

For benchmarking (e.g. comparing a company's losses to the industry mean in a given year), an interpretation of these results is that costs developed using one method should not be compared to costs using other methods. Since most insured companies are not able to use either the Factor-Adjusted or 30-Month methods to calculate costs, a suggested practice is to publish WC costs intended for benchmarking purposes using the Most-Recent method even if costs based on the other methods are also published.

For evaluating trends over time, although trends differed, it is unclear which method is most accurate. An overall issue with evaluating aggregate WC cost trends over time is that WC costs for a given claim continue to increase as the claim matures. This is exhibited in Table 1, where the Most-Recent total incurred values are higher than the 30-Month total incurred values. The 30-Month and Factor-Adjusted methods are both designed in part to address this problem, and in theory should produce a more accurate trend over time. The 30-Month method may be preferred, if only because it is easier to calculate and communicate with insured companies. To accurately determine trends in real costs over time, inflation adjustments should be made, since costs of claims in each accident year are stated in the dollars of different years. Additional research is required to guide the application of inflation factors, given the fact that paid and reserve amounts sum together costs that are paid and expressed in dollars of different years, and the Factor-Adjusted method reserve amounts, unlike reserves as estimated under the other methods, are stated in dollars of future years.

For estimating the absolute magnitude of costs of large groups of claims, it appears necessary to consider Factor-Adjusted costs, because reserves are based on the mean (expected value) of the cost of claims of the same type. We saw that Factor-Adjusted costs are much higher than costs as estimated by the other methods (Table 1), and that trends using Factor-Adjusted cost are also often different from trends based on other methods.

For evaluating intervention effectiveness, the 30-Month and Factor-Adjusted methods may be preferred. Although these two methods produce different trends, the choice between them may be less important in the context of evaluating interventions, since the study design should measure the impact of intervention relative to the 'background' trend. Adjustment for cost trends would necessarily include adjustment for the impact of inflation, as well as any other trends in costs that are independent of intervention.

Conclusions

The differences between valuation methods for WC claims must be understood and communicated to users and audiences before applying for intended uses.

Table 1. Mean and Median Total Incurred Costs per Claim

Year	Number of Included Claims	Number of Excluded Claims	30-Month		Most-Recent		Factor-Adjusted	
			Mean	Median	Mean	Median	Mean	Median
2004	13439	1684 (11.1%)	$6,518	$476	$8,083	$506	$10,287	$496
2005	12304	1652 (11.8%)	$5,763	$516	$6,429	$549	$7,996	$542
2006	10845	1518 (12.3%)	$6,411	$517	$7,445	$560	$10,655	$554
2007	9705	1690 (14.8%)	$5,240	$557	$6,569	$581	$9,455	$582
2008	8358	1606 (16.1%)	$5,964	$553	$6,786	$561	$10,848	$565
2009	6732	929 (12.1%)	$5,654	$624	$5,921	$627	$11,290	$638
TOTAL	61,383	9079 (12.9%)	$5,975	$530	$6,986	$556	$9,948	$552

Figure 1. Total Incurred Geometric Mean Cost per Claim*

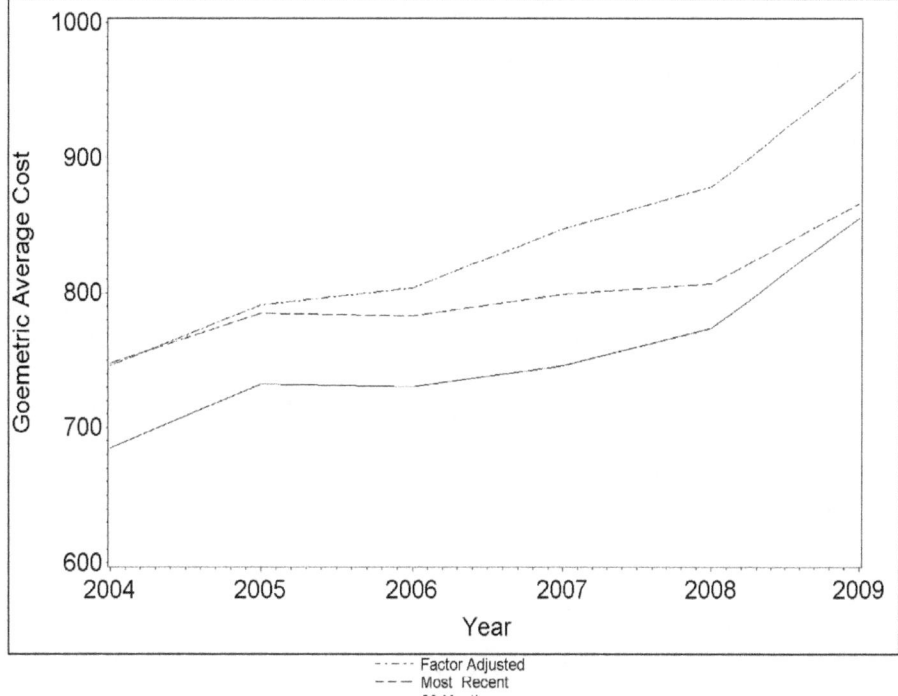

*Factor-Adjusted is a significantly different trend

Figure 2. Total Incurred Mean Cost per Employee*

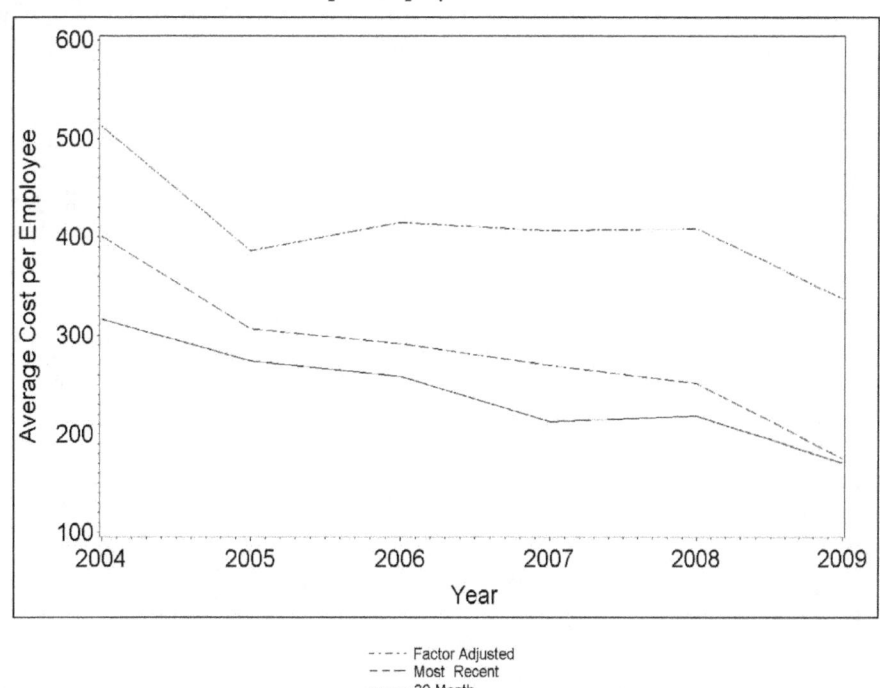

*Trends not significantly different

Development and Evaluation of an Auto-Coding Model for Coding Unstructured Text Data Among Workers' Compensation Claims[1]

Bertke SJ§, Meyers AR§, Wurzelbacher SJ§, Bell J§, Lampl ML*, Robins D*
§National Institute for Occupational Safety and Health, *Ohio Bureau of Workers' Compensation

Introduction

Work-related musculoskeletal disorders caused by ergonomic risk factors (MSDs) such as overexertion and repetitive motion and injuries caused by a slip, trip or fall (STF) are common among workers and result in pain, disability, and substantial cost to workers and employers (Bureau of Labor Statistics, 2011; Liberty Mutual Research Institute for Safety, 2011). The majority of work-related occupational injuries and illnesses can be categorized as a MSD or a STF (Bureau of Labor Statistics, 2011). Improved surveillance of occupational illnesses and injuries (II) classified as MSDs and STFs has been a high national priority, as determined by the National Occupational Research Agenda (NORA). In fact, ninety percent of the time, surveillance of MSDs and STFs were included as strategic goals among the ten NORA sectors' (e.g. manufacturing, construction, wholesale/retail trade [WRT]) agendas. Tracking the incidence and prevalence of MSDs and STFs among Ohio workers is one aim of the partnership between the National Institute for Occupational Safety and Health (NIOSH) and the Ohio Bureau of Workers' Compensation (OBWC).

The OBWC collects claims data primarily to manage claims and determine future workers' compensation premiums. Prior to 2007, OBWC had no systematic way of tracking events or exposures (i.e. causation) such as ergonomic risk factors and slips, trips, or falls. Causation was only recorded in a free-text field (unstructured data) used to describe the work-related cause of the claim. Tracking the incidence and prevalence of MSDs and STFs among Ohio workers would therefore require coding causation for millions of unstructured fields and to do this manually was not feasible.

Recently, Lehto et al (Lehto et al 2009; Wellman et al, 2004) demonstrated that computer learning algorithms using Bayesian methods could auto-code injury narratives into different causation groups, without any manual intervention, efficiently and accurately. The authors demonstrated that the algorithms could code thousands of claims in a matter of minutes or hours with a high degree of accuracy by "learning" from claims previously coded by experts, referred to as a training set. Furthermore, these algorithms provided a score for each claim that reflected the algorithm's confidence in the prediction and, therefore, claims with low confidence scores could be flagged for manual review.

The main goal of this project was to develop and evaluate an auto-coding method which could be used to aid the manual coding of OBWC claim causations as MSD, STF, or other (OTH).

Methods

Case definitions
The case definition for a MSD developed for this study reflected the MSD case definition used by the BLS, which uses the Occupational Injury and Illness Classification System (OIICS) to code nature of injury and event or exposure. The first criteria for MSD cases were those where the nature of injury included sprains, strains, tears; back pain, hurt back; soreness, pain, hurt, except the back; carpal tunnel syndrome;

---

[1] The findings and conclusions in this paper are those of the authors and do not necessarily represent the views of the Ohio Bureau of Workers' Compensation or the National Institute for Occupational Safety and Health.

hernia; or musculoskeletal system and connective tissue diseases and disorders. The second criteria for MSD cases, with few exceptions, were those where the event or exposure leading to the injury or illness was one of the following OIICS codes: bodily reaction (bending, climbing, crawling, reaching, twisting); overexertion; repetition; rubbed or abraded by friction or pressure (contact stress); rubbed or abraded by friction or vibration. Almost all of STF cases were injuries caused by slips, trips and falls, as defined by OIICS. Claims were also coded as a third category, Other (OTH), which included all other II events not classified above as either an MSD or STF. The OTH category included events such as assaults, motor vehicle crashes, contact with objects and equipment, and exposure to harmful substances.

The auto-coding program (described below) was used to identify the causation category of various OBWC claims. For the purposes of this study, causation category was explained by an 'accident narrative' and 'injury category' fields. The unstructured accident narrative is a brief description of how the injury or illness occurred. The most influential field for a manual coder is the accident narrative; however, narratives tend to be noisy, with misspellings, abbreviations, and grammatical errors. For example, a STF narrative reads "IN COOLER, CARRING CRATE TRIP OVER CASE OF BEER HIT CEMENT FLOOR." The structured injury category field was created by OBWC for internal purposes and gives a description of the nature of the injury. It is a categorical field with fifty levels assigned based on the claim's most severe International Classification of Diseases Ninth Revision Clinical Modification (ICD-9 CM) code.

Auto-coding Procedure
The auto-coding procedure developed for this project was based on a process referred to as Naïve Bayes analysis, which is a common text classifier technique (Sebastiani, 2002), and attempted to build upon the work of Lehto et al (2009) in this area. In short, the procedure attempts to calculate the probability a given claim belongs to each possible causation category and the causation category with the highest probability is assigned to the claim. Also, a score value reflecting the probability the claim was coded correctly is assigned. The probabilities are estimated by considering the relevant words of a text narrative and investigating their frequency in the text narratives of all the claims in a training set. For example, the word "FELL" frequently occurs in the narratives of STF claims in the training set and as a result any unknown claim with the word "FELL" in its narrative will be assigned a high probability of being a STF. In addition to considering the accident text narrative, the injury category description field was also considered since, for our study, the definition of an MSD is dependent on how the injury occurred as well as the nature of the resulting injury. Consideration of this additional structured field is an extension of the work of Lehto et al (2009), which only considered the unstructured accident text.

Method of Evaluation
NIOSH evaluated the algorithm on the set of 10 132 un-coded OBWC-insured, single location employers, WRT Sector claims from 2008. To implement our method, NIOSH randomly sampled 2400 claims out of the 10 132 to use as a training set for the algorithm. The claims were randomly sampled evenly across each month and between two claim severity types (lost-time, medical only). Three NIOSH safety and ergonomics experts independently coded each of the 2400 claims as a MSD, a STF, another claim type (OTH), or not otherwise classified (NOC). NOC claims were usually missing an accident narrative or the narrative was too vague to make a determination. Of the 2400 claims, the three coders disagreed on 148 (6.2%) claims and 12 (0.5%) claims were coded as NOC. These 160 claims were removed from the training set resulting in a set containing 2240 manually coded claims.

The auto-coding method was then applied to the remaining 7732 (10 132 minus the 2400 sampled for the training set) un-coded OBWC WRT Sector claims from 2008. As a quality control (QC) measure to evaluate the effectiveness of the algorithm, an additional 800 claims (over 10% of the 7732 un-coded claims) were sampled. These claims were then manually coded by 1 of the three NIOSH experts, blinded to the auto-coded results. The results from the manual coding (which were assumed to be accurate) were then compared to the auto-coded results. The effectiveness of the auto-coding program was measured by the sensitivity, specificity and positive predictive value (PPV).

Results and Discussion
The Naïve Bayes auto-coding program developed in this project took less than 5 minutes to auto-code the 7732 WRT 2008 claims using the 2240 previously coded training set. Table 1 lists the performance of the method in categorizing the 800 randomly sampled QC set into the 3 causation categories. Overall, when using only the text narrative to code claims, the auto-coding method predicted 88.4% of the claims correctly. When the injury narrative code was also considered, there was modest improvement overall (89.9%) in predicting claims. However, there was a large improvement in identifying MSDs, with the sensitivity increasing from 85.4% to 90.3% and the positive predictive value (PPV) increasing from 83.7% to 89.0%. This improvement in identifying MSDs is not surprising since the definition of a MSD depends not only on the cause of the II but also the nature of II.

To investigate how well the score value represents the auto-coding program's accuracy, Figure 1 graphs the percent of claims predicted correctly versus the score value assigned by the auto-coding program. There is a definite trend that claims with lower scores were less likely than claims with higher scores to be coded correctly. However, it appears that the score value tended to slightly overestimate the prediction strength. For example, only 70% of claims with a score between .83 and .85 were coded correctly. Even so, this score can be useful in flagging claims for manual review.

Conclusions
We replicated and expanded upon a Bayesian machine learning auto-coding technique that has been shown to be an effective, accurate and fast technique of identifying the accident causation category for a claim. Our work extended the previous efforts of others in this area by not only considering the accident text narrative, but also the injury category field; these two fields taken together improved the program's overall accuracy. This program will allow us to code many years of OBWC claims data in order to calculate rates of STF and MSD claims by sector and sub-sector. Eventually this benchmarking information will help to target occupational safety and health intervention efforts for Ohio employers. Additionally it will allow researchers to evaluate the effectiveness of injury reduction efforts at larger scales. Similar techniques as described in this paper could be used by other public health practitioners to analyze large sets of existing unstructured text data that is not currently useful.

References
Bureau of Labor Statistics. [2011]. Nonfatal occupational injuries and illnesses requiring days away from work, 2010 Bureau of Labor Statistics News Release: U.S. Department of Labor.

Bureau of Labor Statistics. [2010]. Occupational injury and illness classification manual, V. 2.0. US Department of Labor, Bureau of Labor Statistics, September 2010.

Lehto M, Marucci-Wellman H, Corns H. [2009]. Bayesian methods: a useful tool for classifying injury narratives into cause groups. Injury Prevention, 15: 259–265.

Liberty Mutual Research Institute for Safety. [2011]. 2011 Liberty Mutual Workplace Safety Index (pp. 2). Hopkinton, MA.

Sebastiani F. [2002]. Machine Learning in Automated Text Categorization. ACM Computing Surveys, 34: 1–47.

Wellman HM, Lehto MR, and Sorock GS. [2004]. Computerized coding of injury narrative data from the National Health Interview Survey. Accid Anal Prev, 36: 165–71.

Table 1. Performance statistics of the auto-coding program in classifying claims as STF, MSD or other (OTH)

	N[a]	\multicolumn{4}{c}{Text Only}	\multicolumn{4}{c}{Text + Injury Code}
	N[a]	N[b]	Sensitivity	Specificity	PPV	N[b]	Sensitivity	Specificity	PPV
All Claims	800		88.4%[c]				89.9%[c]		
NOC	6	0	0.0%	100.0%	-	0	0.0%	100.0%	-
MSD	144	147	85.4%	96.3%	83.7%	146	90.3%	97.6%	89.0%
STF	190	205	90.0%	94.4%	83.4%	215	90.5%	93.0%	80.0%
OTH	460	448	89.8%	89.7%	92.2%	439	90.7%	93.5%	95.0%

[a] – Actual number of claims in each causation category
[b] – Number of claims predicted by auto-coding program in each category
[c] – Overall percent of claims coded correctly by the auto-coding program

Figure 1. Graph of percent of claims coded correctly vs. their score value calculated by the auto-coding procedure.

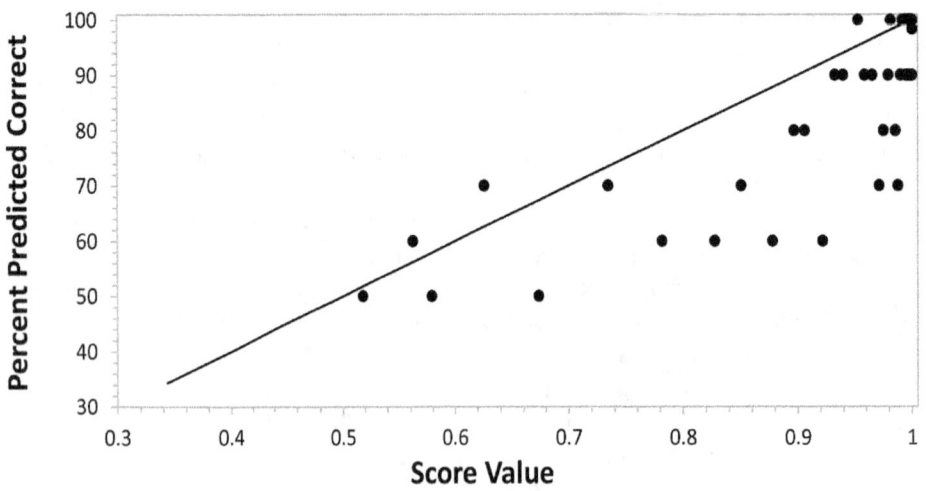

Patterns in Employees' Compensation Appeals Board Decisions: Exploratory Text Mining and Information Extraction[1]

Robin Ackerman, JD, SM§; Janet Ackerman, BA*; Luke Miratrix, PhD#
§U.S. Department of Labor, *Silent Spring Institute, #Harvard University

Introduction

This project applies text mining and basic automated information extraction approaches to thousands of decisions of the Employees' Compensation Appeals Board (ECAB). ECAB considers appeals from determinations of the Office of Workers' Compensation Programs (OWCP) in the U.S. Department of Labor (DOL), which handles compensation claims from federal workers who are injured in the course of employment. EBAC handles as many as 2000 appeals per year; the resulting decisions are publically available through DOL at http://www.dol.gov/ecab/decisions/main.htm.
We considered the following research questions:
(1) Does case outcome vary by issue, condition, year, month, and/or judge?
(2) Can key-phrase extraction algorithms extract meaningful information from ECAB decisions?
(3) What kinds of analyses do text mining and automated information extraction facilitate, with respect to legal decisions?

We examined case outcome by year, month, judges, condition, and issue, among decisions published between 2005 and 2010. We also considered the applicability and utility of novel key-phrase extraction algorithms based on Miratrix et al. (2011), Gawalt et al. (2010), and Ifrim et al. (2008) for these legal decisions.

Methods

ECAB decisions are organized on the DOL website by year and month. We automated the retrieval of all of the links from each of these pages, and downloaded the source code of the pages containing the decisions. Removing HTML tags and other formatting elements yielded the text of the decisions, which we stored in comma delimited text files with year, docket number, and month information (all easily extracted from case URLs). We then pre-processed the decision text to prepare it for analysis with R, an open source software environment and language.

Manual review of decisions from different time periods revealed differences in formatting in certain years. We therefore restricted analysis to cases published from 2005 to 2010, because formatting appeared relatively consistent during this period. Each decision contained a short summary of the central controversy in the case, and this section was labeled "Issue." Case outcomes consistently appeared near the end of each decision as "affirmed," "reversed," and/or "remanded." We classified each decision with two binary variables, the first describing whether causality or work-relatedness of a condition was in question (the Issue), and the second describing whether the applicant had sought compensation for a mental health condition (the Condition), by searching for specific key-phrases. Table 1 summarizes all outcomes and covariates, along with details of how we obtained them. By hand-checking random samples of the decisions, we estimated the sensitivity and specificity of our categorization criteria for Issue and Condition (see Table 2).

After considering crude differences in outcomes between case categories, we performed logistic regression, regressing case outcome onto all covariates, to examine associations between the case outcome and judge, condition, issue, year, and month.

---

[1] The views expressed in this article are the personal views of the authors and do not purport to represent the official views of the Department of Labor. The findings presented here do not represent findings of the Department of Labor.

Separately, we used sparse-regression techniques to identify words and phrases that differentially appeared in the ECAB decisions that involved both mental health conditions and causality/work-relatedness, as compared to the other, baseline, decisions. These techniques are described further in Miratrix et al. (2011), although we here extended them to allow for variable-length phrases, using methods similar to those described in Ifrim et al. (2008). These methods return phrases that are able to predict (i.e., are substantially correlated with) condition type, adjusted for overall phrase prevalence. Thus, in comparing categories of legal decisions, formulaic legal phrases that generally appear in both categories will be excluded. Ideally, phrases that meaningfully differentiate a class of cases from the others will be highlighted.

Results

Table 2 presents estimated sensitivity and specificity values for the markers used to identify conditions (mental health conditions vs. all others) and issues (causality/work-relatedness vs. all others). Two analyses are shown, one for the emotional condition indicator and one for the causality/work-relatedness indicator. The total numbers of decisions labeled as positive or negative are shown under the "Total" column. For each analysis, we hand-checked 100 positively labeled decisions and 100 negatively labeled decisions. Results are presented under the "% Cor" column. We then used these samples to estimate total performance of the labeling schemes, and calculated sensitivity and specificity from these estimates.

Table 3 displays results from our final logistic regression of case outcome onto the covariates, after testing individual covariates with likelihood ratio tests. Our models revealed no remarkable or highly significant associations between case outcome and any individual judge. However, likelihood ratio tests indicate substantial improvement in model fit with the inclusion of the judges (Chi-squared = 23.4, 7 degrees of freedom, p value =0.001). OWCP determinations were affirmed more often in cases involving an emotional condition, as compared to those that did not, and more often in cases where the causality or work-relatedness of a condition or disability was at issue, as compared to cases where causality was not at issue. No clear association emerged between outcome and month or year.

Table 1. Variable Definitions

Outcome	A binary variable differentiating cases in which ECAB entirely affirmed OWCP's determination from those in which ECAB at least partially reversed OWCP's findings or remanded the case.
Issue	A binary variable distinguishing cases that centered on the causality of a condition or disability from all other cases, using marker phrases "caus" (except when appearing within the word "because") and/or "performance of duty."
Condition	A binary variable differentiating cases in which the appellant originally sought compensation for a mental health condition from all other cases. This was accomplished by searching for the marker phrase "emotional condition" within each entire decision text.
Judges	A binary variable for each judge who served on ECAB between 2005 and 2010.
Month and Year	Values for month and year were extracted from case urls.

Table 2. Sensitivity and Specificity of Markers for Condition and Issue

Labeling	Estimated					
	Total	% Cor (sample)	Positive	Negative	Sensitivity	Specificity
"Emotional Condition"	1479	94/100	1390	89	0.83	0.99
No "Emotional Condition"	9735	97/100	292	9443		
Total Cases	11214					
"Causality/ Work-Relatedness"	4236	99/100	4194	42	0.75	0.99
No "Causality/Work-Relatedness"	6978	80/100	1396	5582		
Total Cases	11214					

Table 3. Logistic Regression on Case Outcome

	Estimate	Std. Error	P(>|t|)	Odds Ratio	95% Confidence Interval
Intercept	0.48	0.31	0.123	1.64	0.89, 3.04
Condition = Emotional	0.19	0.07	< 0.01	1.19	1.03, 1.37
Causality at Issue	0.78	0.05	< 0.01	2.18	1.98, 2.40
Judge Gerson	0.13	0.11	0.241	1.15	0.92, 1.43
Judge Groom	0.20	0.11	0.069	1.23	0.98, 1.54
Judge Haynes	0.08	0.11	0.431	1.09	0.88, 1.34
Judge Kiko	0.12	0.11	0.259	1.14	0.91, 1.40
Judge Koromila	0.01	0.11	0.910	1.02	0.81, 1.26
Judge Kanjorski	-0.01	0.16	0.966	0.99	0.73, 1.35
Judge Thomas	-0.19	0.13	0.154	0.83	0.64, 1.07

The key-phrase extraction algorithms identified the following words and phrases as being characteristic of cases where both an emotional condition and causality/work-relatedness were at issue, as compared to all other cases:
illness that is
Lillian Cutler
imposed by
reaction to
depression
anxiety
psychiatrist
causally related
the truth
factual evidence identifying
incidents alleged to
allegations of
are alleged
a factor of

Many of these words and phrases appear within boilerplate passages of ECAB decisions that establish basic principles and rules, as is evident in the following excerpts:
Workers' compensation law does not apply to each and every injury or illness that is somehow related to a claimant's employment. In the case of Lillian Cutler, the Board explained that there are distinctions as to the type of employment situations giving rise to a compensable emotional condition under FECA. [C.E., Docket No. 10-461 (issued November 23, 2010)] (emphasis added).
To establish that an emotional condition arose in the performance of duty, a claimant must submit the following: (1) medical evidence establishing that she has an emotional or psychiatric disorder; (2) factual evidence identifying employment factors or incidents alleged to have caused or contributed to the condition; and (3) rationalized medical opinion evidence establishing that the emotional condition is causally related to the identified compensable employment factors. [T.G., 58 ECAB 189 (2006)] (emphasis added).

Discussion
Automated text extraction techniques facilitate the study of large numbers of legal decisions, where manual processing is impractical. Our analyses emphasize this point. However, both logistic and ordinary least squares regression models rely on our classification of issues and conditions; the estimated sensitivity and specificity of our markers reveal some misclassification, and alternate categorizations may be more useful or meaningful, depending on the precise questions of interest. Additionally, important predictors of outcome may be missing from these models. We report associations but draw no conclusions about causation. Nonetheless, our model results are largely unsurprising: the outcomes of decisions involving emotional conditions or causality are especially likely to be consistent with earlier determinations. We did not identify any particular judge as an outlier in relation to case outcomes, but likelihood ratio tests indicated significant improvement in model fit when judges were included, suggesting that these differences are non-negligible and that judges do vary.

The Sensitivity scores (see Table 2) are moderately low due to a combined impact of the greater prevalence of negatively marked cases, which is to be expected when comparing a specific case type of interest to a more general baseline, and conservative labeling that does not mark cases without specifically identified key-phrases as positive. However, this is possibly of minor impact, especially in the key-phrase extraction, in that it will dilute the difference in appearance rates between phrases in the positive and negative examples, rendering it more difficult to find them, but it is unlikely to introduce artifacts into the final summaries. A similar argument applies to the regression models as well. Of course, it is worth seeking superior labeling methods in future work.

The key-phrase extraction algorithm can reveal differences in boilerplate language and citations between, for example, cases involving both causality/work-relatedness and an emotional condition as compared to cases without these characteristics. These phrases seem to highlight points of law that are characteristic of these cases, indicating that such techniques may facilitate the development of automated case content analysis, and may even aid in the development of refined legal taxonomies.

References

1. Gawalt, B., J. Jia, L. W. Miratrix, L. Ghaoui, B. Yu, and S. Clavier (2010). Discovering word associations in news media via feature selection and sparse classification. In MIR'10, Proceedings of the International Conference on Multimedia Information Retrieval, Philadelphia, Pennsylvania, USA, pp. 211–220.

2. Ifrim, G., G. Bakir, and G. Weikum (2008). Fast logistic regression for text categorization with variable-length n-grams. In 14th ACM SIGKDD International Conference on Knowledge Discovery and Data Mining, New York, NY, USA, pp. 354–362. ACM.

3. Miratrix, L.W., Jia, J., Gawalt, B., Yu, B., El Ghaoui, L. (2011). What is in the news on a subject: automatic and sparse summarization of large document corpora. UC Berkeley Dept. of Statistics Technical Report #801.

4. R Core Team (2012). R: A language and environment for statistical computing. R Foundation for Statistical Computing, Vienna, Austria. ISBN 3-900051-07-0, URL http://www.R-project.org/.

Identifying Workers' Compensation as the Expected Payer in Emergency Department Medical Records[1]

Larry L. Jackson, PhD, Susan J. Derk, MA, Suzanne M. Marsh, MPA, Audrey A. Reichard, OTR, MPH
National Institute for Occupational Safety and Health

Introduction

The National Institute for Occupational Safety and Health (NIOSH) uses the National Electronic Injury Surveillance System—occupational supplement (NEISS-Work), an emergency department (ED) based surveillance system, to produce national estimates of nonfatal occupational injuries and illnesses treated in US hospital EDs [Derk, 2007]. The occupational injury and illness data are collected from a national, probability-based, stratified cluster sample of 67 hospital EDs. At each hospital, abstractors review ED medical records for work-related injuries and illnesses. The abstractors identify cases as work-related if the injuries or illnesses occurred while the patient was doing work for compensation, work or chores related to agricultural production, or work conducted as a volunteer for an organized group. Civilian noninstitutionalized workers without regard to employment arrangement, worker status, age, industry, or business size are included. We use a work-related case definition similar to Occupational Safety and Health Administration (OSHA) Recordkeeping Rules [OSHA, 2012]. Hence, we include injuries and illnesses caused or significantly made worse by work for which the patient was seen in the ED. We presume that being seen in the ED qualifies as medical treatment beyond first aid. NEISS-Work does not capture the medical treatment provided. We exclude common illnesses and selected medical conditions along with cases involving self-medication, alcohol & drug cases, drug screening, and second visits to the ED. Hospital record abstractors review the entire medical record and use narrative and coded information in the ED record to assess the work-relatedness of each case. Workers' compensation insurance as the expected payer, by itself, is sufficient to identify a case as work-related, providing that the narrative information in the medical record does not contradict the injury/illness as being work-related. However, workers' compensation insurance is not required for a case to be identified as work-related. Narrative information in the registrar's, nurse's, and/or doctor's notes indicating that an injury/illness occurred at work is sufficient to identify a case as work-related.

To better understand issues in identifying work-related injuries and illnesses in ED medical records, we audited records at about one-third of the hospitals participating in NEISS-Work. The primary goal of the audits was to estimate the number of work-related cases missed or non-work cases misclassified as work-related. A secondary goal was to better understand the various issues that abstractors deal with in identifying work-related cases on a daily basis. This report summarizes some of the issues that we identified qualitatively and in particular, the ability to identify workers' compensation (WC) as the expected payer for the medical care.

Methods

Abstractors collect NEISS-Work surveillance data at each hospital on a daily basis throughout the year. We use these data to provide national estimates of the number and rate of occupational injuries and illnesses treated in EDs. The surveillance data collected include the characteristics of the injured/ill worker,

---

[1] The findings and conclusions in this report are those of the authors and do not necessarily represent the views of the National Institute for Occupational Safety and Health

nature of injury/illness, and the event or circumstances leading to the ED visit. In addition, some employment related information and the expected medical payer are collected. For work-related cases, the hospital abstractors identify the expected payer at the time of data abstraction based on administrative information in the ED record. Abstraction is typically done within 1 day to 1 week of treatment. The primary expected payer at time of admission may change as the billing process proceeds (e.g., between initial abstraction and audit). Moreover, the expected payer may not represent the final source(s) of medical payment. The NEISS-Work expected payer categories include: Injured/ill worker—personal insurance or self-pay; Employer/union—private insurance or direct-pay; Private health insurance—unspecified policy holder; Workers' Compensation; Other government; Other; and Not stated or unknown.

As part of our ongoing efforts to improve the NEISS-Work surveillance and understand its limitations, we conducted audits in 20 NEISS-Work hospitals stratified by hospital size (number of ED visits/year) and geographically distributed across the U.S. At each hospital we examined all ED records for a specified number of treatment days such that the total number of patient records reviewed exceeded 1,000 cases. Treatment periods reviewed typically ranged from a week for large hospitals up to three months for small hospitals. We abstracted all work-related cases at each hospital (~3% of all cases) for the time period reviewed. Although not conducted as a quantitative review, we observed numerous electronic medical record, charting, and other data issues while doing the audits.

Results

Preliminary, routine surveillance data for nonfatal occupational injuries and illnesses treated in a U.S. hospital ED in 2011 indicate that WC was the expected payer in about 55% of the estimated 2.9 million occupational injuries and illnesses (Figure 1). Cases listed as Government, non-entitlement programs (~2%) and employer/union (~2%) may also have been WC insurance. For more than one-fourth of the cases, personal health insurance or self-pay was the expected payer.

During our audits, we found that record abstractors rely on a multitude of information systems that vary widely across the NEISS-Work hospitals. All audited hospitals had an electronic ED registration system. Some EDs used fully electronic medical record systems. Some hospitals used a combination of electronic and paper records. A few hospitals used paper charting processes only. Hospitals used as many as 4 independent electronic record systems with diverse levels of integration, technologies, sophistication, and accessibility. Because of hospital customization and requirements, electronic medical record systems varied across hospitals even when provided by the same information system vendor.

In general, NEISS-Work record abstractors had access to select employment and insurance information which included the guarantor, employer name, and insurance details for one or more insurers (Table 1). The prevalence and completeness of numerous fields and forms/reports along with conflicting information in various components of the medical records varied greatly between hospitals.
Our audits suggested that to identify work-related cases in the ED medical record it is critical to have narrative information indicating "at work" in the doctor's or nurse's notes or standardized "at work" check boxes that are actually used. Because the NEISS-occupational supplement is the only NEISS program collecting expected payer information it was unclear if some abstractors only reviewed the expected payer information when other medical record information already indicated that the case was work-related. A few hospitals appeared to be indicating WC as the expected payer for all work-related cases as a matter of practice. Common problems in identifying the patient as employed and the expected payer as WC are highlighted in Table 2. Also, abstraction of the expected payer information within 1 to 2 days of treatment occasionally differed from the expected payer indicated during the audits.

Summary

NEISS-Work captures occupational injuries and illnesses for all types of civilian workers including the self-employed and family members working in a family business. The latter types of worker are often not covered by WC. Nevertheless, approximately 90% of U.S. civilian workers are legally required to be covered by WC.[2] Routine NEISS-Work surveillance data suggest that 55% of the ED-treated work-related cases in 2011 had an expected payer of private employer workers' compensation insurance. An additional 4% may have had WC coverage through government agencies or direct, self-insured employer payments. Our qualitative information from the audits identified many issues in capturing WC as the expected payer, but provided no clue as to whether WC is being underutilized in the ED, misidentified, or simply not identified in the medical records. Obviously, the ultimate proportion of the ED-treated cases paid by WC is unknown and may vary widely from hospital to hospital and state to state. In a 10-state telephone interview study, the proportion of workers who self-reported a work-related injury with payment by WC varied from 50-77% [Bonauto et al., 2010]. The survey results included all forms of medical treatment, not just occupational injuries and illnesses treated in an ED as reported here. Utilization of WC may vary across medical venues.

To improve the capture of WC as the expected payer in NEISS-Work ED surveillance will require specialized training for abstractors, largely on an individual hospital basis. Although potentially feasible, it is unclear if the training would significantly influence or change the proportion of cases indicated with WC as the expected payer. Improvements and standardization of insurance classification practices from state to state and within electronic medical record systems would aid identification of WC as well as other insurers as the expected payer in a broad spectrum of medical venues. The standardization would likely aid surveillance and injury prevention activities. Ultimately, knowing how many injured/ill workers actually file a WC claim and the claim is paid compared to the number of cases with WC indicated as the "expected payer" would further aid interpretation of our surveillance data.

References

BLS [2012]. Labor force statistics from the Current Population Survey. Available at http://www.bls.gov/cps/tables.htm.

Bonauto DK and others [2010]. Proportion of Workers Who Were Work-Injured and Payment by Workers' Compensation Systems --- 10 States, 2007. MMWR July 30, 2010 / 59(29);897-900.

Derk SJ, Marsh SM, Jackson LL. [2007]. Nonfatal Occupational Injuries and Illnesses— United States, 2004. MMWR April 27, 2007 / 56(16);393-397.

OSHA [2012]. OSHA Injury and Illness Recordkeeping. Available at http://www.osha.gov/recordkeeping/.

Sengupta I, Reno V, Burton JF Jr, Baldwin M [2012]. Workers' Compensation: Benefits, Coverage, and Costs, 2010. National Academy of Social Insurance, Washington, DC. 120 p.

---

[2] Percentage was derived from the number of covered workers (124.5 million) [Sengupta et al., 2012] and the number of U.S. employed workers (139.1 million) in 2010 [BLS, 2012].

Figure 1. Percentages of expected payers for nonfatal occupational injuries and illnesses treated in U.S. hospital emergency departments—2011 (preliminary data).

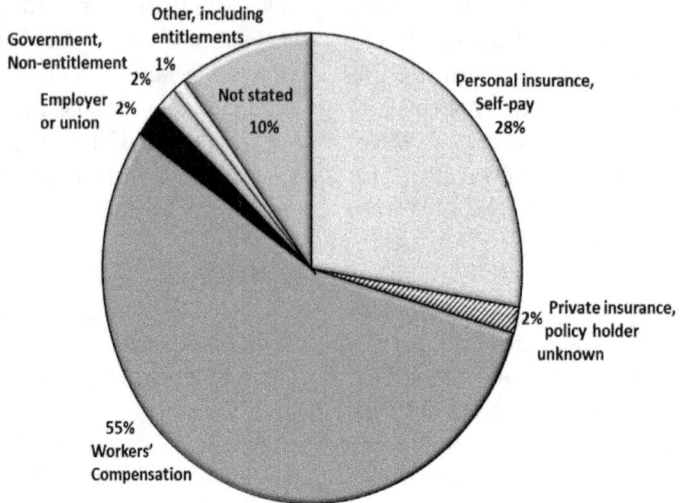

Table 1. Qualitative assessment of prevalence of employment and insurance information fields or forms in medical records and the completeness of the information.

Prevalence*	Information Field or Form	Description
Common	Guarantor	The person or entity that is financially responsible for patient.
Common	Employer name (patient's and/or guarantor's employer)	The employer's (or union, in some instances) name and address.
Moderate	Patient's occupation	Free text that may contain employment status terms in place of occupation (e.g., self-employed) or non-employment terms (e.g., retired, student, homemaker)
Rare	Employment status	Uniform Billing (UB04) classes: Employed (full- or part-time), Homemaker, Child, Active Military Duty, Retired, Self-employed, Student, Unemployed, Disabled, or Unknown.
Common	Primary, secondary, and tertiary insurance providers	Provider/carrier name, plan type (e.g., HMO, PPO, Medicare), policy #, group #, subscriber, etc.
Rare	Workers' Compensation	Carrier name and plan information.
Rare	First report of injury form	Narrative description of employment, injury circumstances, and injury characteristics completed at the ED.
Rare	Insurance verification form	Hospital verification of insurance and/or employment/work-relatedness of incident.
Common	Financial class	1-3 digit grouping of insurance types (e.g., BCBS, Medicare, Medicaid, Commercial, Workers' Compensation, and Military). Used for high level reporting. Transparent to registration and billing users.
Rare	Other	Registrar or nurse notes "*Gave patient Workers' Compensation forms*"

*Prevalence: Common = common field and usually completed; Moderate = field often available, but only moderately completed; Rare = field, form, and/or information rarely available.

Table 2. Common employment and expected payer identification issues in medical records.

Data elements	Issues
Employer	Frequently provided, but may not be correct business name; employer may be missing, incomplete, not collected, or not updated from a prior visit
Occupation	Often missing, inaccurate response, not collected, or not updated from a prior visit
Industry	Never collected, except indirectly in healthcare provider notes (e.g., "hurt at meat packing plant")
Accident location	Rarely explicitly specified except for motor vehicle incidents; often implied in healthcare provider notes (e.g., "hurt at work" & employer is a restaurant)
Injury at work	Field is infrequently available. When the field is present, the default response is commonly equal to "no"
Insurance	Abstractors require local knowledge of insurance providers and plan types; Workers' Compensation carriers may not be obvious
Financial class	Knowledge of codes is essential; however, financial class may not match insurance plans listed
Employment and employment status indicators	May not be updated for current visit; may default to child or student for young workers or retired for older workers

Utilizing Workers' Compensation Data to Evaluate Interventions and Develop Business Cases [1]

James W. Collins, PhD, MSME, Jennifer L. Bell, PhD
National Institute for Occupational Safety and Health

Introduction

Workers' compensation statutes have been enacted in all 50 United States and Washington D.C. to provide compensation for injured employees to pay for lost wages, medical costs, and rehabilitation for persons who become ill or injured as a result of their employment. Data collected through the workers' compensation process are not only used to manage work-related injury claims, but are also used in aggregate with other data to develop prevention strategies and to manage work disability. This paper describes two NIOSH studies that used workers' compensation data to: 1) conduct descriptive epidemiologic analyses to characterize the injury problem, 2) benchmark baseline injury rates, 3) develop injury prevention strategies, 4) evaluate the effectiveness of prevention strategies in rigorous intervention trials, 5) develop a business case based on the economics of the injury problem and the intervention, 6) contribute to the evidence-base of science supporting interventions addressing the leading causes of workers' compensation in the Health Care Industry, and, 7) market the research findings to key opinion leaders, policy makers, educators and practitioners to gain support for implementation of "best practices", curriculum development, standards development and legislative initiatives to stimulate industry wide implementation of the prevention programs (Collins, Bell & Grönqvist 2010).

Methods

This paper summarizes two intervention trial studies in health care workers (Collins et al., 2004; Bell et al., 2008). The first study was a nine-year intervention trial that evaluated a "best practices" safe patient handling injury prevention program in a dynamic cohort of nursing personnel (Collins et al., 2004). The intervention was implemented in six nursing homes. Injury rates, costs, and lost and restricted workdays were compared for the three-year pre-intervention period (1995-1997) and the six-year post-intervention period (1998-2003).

In the second study, a multidisciplinary research team (Bell et al., 2008) designed, implemented, and evaluated a comprehensive "best practices" STF prevention program in three hospitals based on the findings from an analysis of historical worker injury data, hazard assessments, and laboratory studies. The field study, conducted in conjunction with a hospital corporation, examined the injury experience of a cohort of approximately 17,000 hospital staff for a 10-year period from 1996–2005.

Both studies were pre-post long-term intervention trials that evaluated "best practices" injury prevention programs in large cohorts of health care workers to identify ways to effectively reduce exposures, identify "best practices," and to evaluate intervention effectiveness in real world settings. The initial analysis of workers' compensation injury claims data was conducted for each study to examine the injury problem. Narrative information from the injury report was used to code the injuries and identify injury events that could be targeted for intervention.

As a precursor to the two field studies described in this paper, laboratory studies were

---

[1] The findings and conclusions in this report are those of the authors and do not necessarily represent the views of the National Institute for Occupational Safety and Health

conducted to examine: 1) reductions in the physical exposures associated with proposed patient lifting interventions (Zhuang et al., 1999; Zhuang et al., 2000) and, 2) the friction characteristics of promising slip-resistant shoes and hospital flooring surfaces when dry and contaminated with water, oil, and cleaning solutions (Collins, Bell, and Grönqvist 2010).

Results

This paper describes the process of how workers' compensation data were used to conduct comprehensive multi-disciplinary research studies addressing the leading causes of workers' compensation claims among health care workers. In the SPH nursing home study, there was a 61% reduction in the workers' compensation claims rate (adjusted rate ratio (0.39, 95% CI: 0.29-0.55) and a 66% reduction in the lost workday injury rate (rate ratio=0.34, 95% CI: 0.20-0.60) for musculoskeletal injuries associated with resident handling after the intervention was introduced. A business case for safe patient handling programs was developed by determining that the initial investment ($158,566) for lifting equipment and worker training was recovered in less than three years based on post-intervention annual savings ($55,000) in direct medical and indemnity workers' compensation costs. Because, this health care system was self-insured, workers' compensation savings were realized immediately. In the hospital STF study, the STF workers' compensation claims rate declined by 59% (adjusted rate ratio (0.41, 95% CI: 0.33-0.54; 1.66 claims per 100 FTE to 0.76 claims per 100 workers per year). Due to the need for brevity in this paper and the extensive findings from these two research studies, readers are referred to the main papers resulting from these studies (Collins et al., 2004; Bell et al., 2008; Collins, Bell, & Grönqvist, 2010) for additional details. Study findings were also incorporated into user-friendly documents designed to facilitate implementation of safety measures in the workforce (Collins et al. 2006, Bell et al. 2010).

Strengths and Limitations of Using Workers' Compensation Data in Research Studies

There are at least two strengths inherent to the use of workers' compensation data. The first is the minimal cost to obtain workers' compensation data. Because workers' compensation records are routinely collected for purposes other than research, the cost of obtaining workers' compensation data is not expensive. Second, is the ability to examine the economic factors associated with the intervention. When the cost to implement the intervention can be shown to be less than the savings in reduced medical and indemnity expenses, those who make decisions about funding prevention programs can more easily justify allocating resources to make changes.

As a research tool, workers' compensation data are not without limitations. The narrative text describing the injury causing event can lack important specificity about the circumstances of an event. In the two NIOSH studies, the narrative text from other injury data systems (first reports of injury, occupational health nurse logs, and OSHA logs) was used to provide additional information on the nature of the injury and circumstances of the incident. Another potential limitation to using workers' compensation data is underreporting; we don't know if everyone who was injured at work filed a workers' compensation claim. People who were injured at work but treated outside the system would be missed.

A limitation in the approach used to create the business case in the safe resident lifting study was that the business case did not consider indirect costs and was based on direct costs only and did not attempt to estimate indirect costs. The reductions in direct costs were so substantial that the original capital expenses to purchase mechanical lifting equipment and provide worker training were recovered in slightly less than three years. The return on investment would have been shorter if savings in indirect costs were considered (for example, lost wages, cost of hiring and training replacement workers, etc.).

Discussion

Although workers' compensation systems are not designed for primary prevention, these two studies demonstrate how workers' compensation data can be utilized to identify significant worker injury problems and to design, implement and evaluate "best practice" injury prevention programs. The analysis of the workers' compensation data also informed our decisions about what interventions to implement, in what populations, and how they should be implemented.

Intervention effectiveness studies often report outcomes measures such as rate ratios, confidence intervals, p-values, and statistical significance to make inferences about the impact of interventions on injury rates. A background in statistics is generally required for meaningful interpretations of these outcome measures. One of the distinct advantages of using the cost data that can be obtained from workers' compensation records is the ability to report on the cost implications of the intervention. The direct costs associated with return on investment and business case findings based on intervention costs balanced with medical and indemnity expenses can be readily interpreted by most people. Pre- and post-intervention medical and indemnity expenses can be compared to estimated savings in direct costs attributed to the intervention. This information can be used to develop business cases that have meaning to those who are making decisions about which prevention programs their company should invest. Corporate leadership has an interest in protecting workers, but when a business case can be presented to management that demonstrates a prevention program significantly reduced worker injuries and also paid for itself, this sends a powerful message that is likely to lead to replication of the prevention program in other settings.

The objective of this paper was to describe the use of workers' compensation data as part of two intervention trials that provided practical information for owners of healthcare facilities, administrators, nurse managers, and safety and health professionals who are interested in replicating these types of programs in their facilities. The research demonstrated that "best practices" safe patient lifting and slip, trip, and fall prevention programs decrease caregiver injuries, lost workdays, and workers' compensation costs and improves employee recruitment and retention, employee morale, and quality of care for residents (Collins et al., 2004; Nelson et. al., 2008; Bell et al., 2008). Using workers' compensation data as part of a comprehensive evaluation of prevention programs makes good business sense and can help inform decisions about reducing worker risk. The hospital corporation participating in this research was self-insured so the cost of insurance premiums was a not an issue in these studies: if an injury was prevented, savings in workers' compensation costs were realized immediately. It is hoped that the evidence-base of science demonstrating the effectiveness of STF and patient lifting prevention programs will facilitate widespread replication of these types of programs in other healthcare facilities, leading ultimately to national declines in the leading causes of work-related injuries among health care workers.

References

Bell JL, Collins JW, Dalsey E, Sublet V (2010). Slip, Trip, and Fall Prevention for Healthcare Workers. DHHS (NIOSH) Publication No. 2011-123. http://www.cdc.gov/niosh/docs/2011-123/pdfs/2011-123.pdf

Bell JL, Collins JW, Wolf L, Grönqvist RA, Chiou S, Chang W-R, Sorock GS, Courtney TK, Lombardi DA, Evanoff B (2008). Evaluation of a Comprehensive Slip, Trip, and Fall Prevention Programme for Hospital Employees. *Ergonomics* 51(12):1906-1925.

Bureau of Labor Statistics (BLS). 2011. Table R8. Incidence rates for nonfatal occupational injuries and illnesses involving days away from work per 10,000 full-time workers by industry and selected events or exposures leading to injury or illness, 2010.

Bureau of Labor Statistics, U.S. Department of Labor, Washington, DC. Survey of Occupational Injuries and Illnesses in cooperation with participating State agencies. Accessed October 2, 2012 at: www.bls.gov/iif/oshwc/osh/case/ostb2832.txt

Collins JW, Nelson A, Sublet V (2006). Safe Lifting and Movement of Nursing Home Residents. DHHS (NIOSH) Publication No. 2006-117. http://www.cdc.gov/niosh/docs/2006-117/

Collins JW, Bell, JL, and Grönqvist, R (2010). Developing Evidence-Based Interventions to Address the Leading Causes of Workers' Compensation Among Healthcare Workers. *Rehabilitation Nursing,* 225-235.

Collins, JW and Bell, JL (2010). Translating Injury Prevention Research into Workplace Practice. Conference Proceedings, Keynote address for the 46th Annual Conference for the Australian Human Factors and Ergonomics Society, pp. 1-10.

Collins JW, Wolf LD, Bell J and Evanoff, B. (2004). An Evaluation of a "Best Practices" Back Injury Prevention Program in Nursing Homes. *Injury Prevention*, 10;206-211.

Nelson A, Collins JW, Siddharthan K, Matz M, Waters T (2008). Link Between Safe Patient Handling and Patient Outcomes in Long-Term Care. *Rehabilitation Nursing,* 33, 1;33-43.

Zhuang Z, Stobbe TJ, Collins JW, and Hsiao H. "Psychophysical Evaluation of Assistive Devices for Transferring Residents." *Applied Ergonomics*, 31 (2000) 35-44.

Zhuang Z, Stobbe TJ, Hsiao H, and Collins JW. "Biomechanical Evaluation of Assistive Devices for Transferring Residents." *Applied Ergonomics*, 30 (1999) 285-294.

Gender, Age, and Risk of Injury in the Workplace

Frank Neuhauser, MPP, Anita K. Mathur, Ph.D., Joshua Pines, BS
University of California, Berkeley, Center for the Study of Social Insurance

Introduction

Over the past half century women have been increasing their participation in the workforce and increasing their representation in traditionally male occupations like construction, manufacturing and transportation. [1,2] This tide of female employment has been accompanied by an increase in the fraction of women among those injured in the workplace, (21% in 1977, to 33% in the early 1990s and 39% in 2009). [3,4] According to BLS, over 360,000 women in the US suffered work injuries resulting in days away from work in 2009.

Despite these trends, our knowledge of how occupational injury risks for women might differ from men is very limited. In addition, we know almost nothing about how these risks interact with age. Older workers comprise a growing fraction of the workforce because the population is aging[5], Social Security retirement age is increasing[6], and many older workers are choosing to stay in the labor force for economic reasons[7].

The research on the impact of gender and age on occupational injury incidence has been hampered by data limitations. Job risk is the interaction of worker characteristics with the specific risks of a specific job. Job risk ideally will be defined along both dimensions of industry and occupation. However, until now no datasets covering a broad range of occupations and industries have had sufficient detail on job risk and hours of exposure to allow researchers to jointly analyze the effects of gender and age on injury rates.

This paper uses a unique combination of data and a new cross-walk for industry and occupation variables to fill many of the gaps in our knowledge. For the first time we are able compare injury incidence by gender and age while simultaneously controlling for both occupation, industry, and hours worked across the full range of workers and jobs.

Data & Methodology

Our approach is to compare the actual injury incidence by gender for various age ranges with the expected incidence based on job risk and hours of exposure. To do so, we created a unique dataset by merging data from the Current Population Survey (CPS), workers' compensation insurance rates by job classification from the California Department of Insurance (CDI) and California injury and illness data from the Workers' Compensation Information System (WCIS) maintained by the California Division of Workers' Compensation.

The CPS data are from the 2003-2008 "Earners Study" sub-sample of California households. CPS households are interviewed for a total of 8 months; information on earnings is collected at the 4th and 8th interview (known as the "outgoing rotations"). CPS data on work status include identification of self-employment. We exclude the self-employed because they are not covered by workers' compensation. We had a pair of 3-digit codes available for each worker to define industry and occupation of their primary job, resulting in approximately 10,000 combinations along the two dimensions.

We define the relative risk of injury for each worker by linking workers' compensation insurance premium rates for each individual worker in the CPS based on the 3-digit industry-occupation pair. Workers' compensation insurance has a unique coding system, called "class codes," that sorts jobs according to similar risk levels and assigns a "manual" premium rate to each class. The manual premium rate is based on the system-wide underlying claim costs relative to payroll in each class code. The "California Workers' Compensation Uniform Statistical Reporting Plan", published by the Workers' Compensation Rating Bureau of California, gives a detailed description of the classes used by California[8]. The class coding in California is very similar to that used in nearly all other states. Frequently the groupings cross industry and occupation categories when the risks are considered similar. Clerical and professional occupations are generally grouped independent of industry, a secretaries at a construction firms are grouped with a secretaries at an accounting firms. On the other hand, a nurse will be coded differently if he or she works in a hospital versus a doctor's office. The level of discrimination in workers' compensation codes is quite fine, with about 500 different classifications. Premium rates across class codes differ by a factor of 100 or more between the highest and lowest risk classes.

Our focus is on the relative frequency of injuries across different groups. Premium rates are not a perfect proxy for relative frequency, since they are a combination of differences in both frequency and cost of claims across classes. However, in practice, frequency dominates as a driver of differences in premium between classes. The correlation between frequency and premium, weighted by payroll is .793 (p<.001), while the weighted correlation between cost per claim and premium is .046 (p<.01).

For a previous study,[9] UC Berkeley developed a cross-walk between each of the approximately 10,000 industry-occupation pairs in the CPS and the related class codes used by workers' compensation insurance. Using this crosswalk allowed us to link the risk value for a class code to each worker in the CPS.

The risk value we use is the workers' compensation premium rate calculated for the specific class code. The manual premium rate for each class is published by California's Department of Insurance. We use the workers' compensation premium rates for the mid-point year (2005) in our CPS sample to calculate a single risk value for each class code. We do this because the level of premium rates can change a great deal year-to-year in response to law, regulation, and insurance market changes. The relative rates between classes, however, change much more slowly. And, for this work, we are only interested in the relative risk between classes.

Workers' compensation premium rates are published as (manual premium)/($100 of payroll). Our interest is calculating a relative value per standard unit of worker exposure to risk. To standardize payroll into exposure units using the CPS, we calculated the average hourly wage among all workers in CPS in each workers' compensation class code. Then we divided by $100 by the average wage to estimate average hours of exposure represented by $100 of payroll. Finally, we divide through the premium rate by the number of hours represented by $100 of payroll. This gives us standard unit of risk/hour for each worker in our California-CPS sample. More simply:

$premium/$100*$100/(Average hourly wage) = Risk.

For example, if the average hourly wage for workers in a class, weighted by the hours worked is $20/hour and the premium rate is $5/($100 payroll), the relative risk per hour is ($5/$100*$100/$/20) = 1.0. Issues related to generally higher wages for older workers or for men versus women in the same job do not enter into this specification once the average hourly wage is calculated for all workers in the class code. Also, the absolute value of the standard units do not have a meaning, only the relative risk of an hour of work exposure between workers or groups of workers.

Using these relative risk values for each industry-occupation pair in the CPS and the number of hours worked in the past week reported by CPS respondents, we calculate a risk value for each worker in our CPS sample. Then we create the expected distribution of injuries for each cell in a 16-cell table that divides respondents by gender and eight age ranges. The expected distribution is the fraction of all occupational conditions that is expected to fall in each gender and age cell based on the fraction of workers in the cell, their hours worked, and the relative risk of their occupations.

We obtained data on the actual occupational injury and illness distribution in California based on all cases reported to the California Division of Workers' Compensation (DWC) Workers' Compensation Information System (WCIS) for the years 2002-2008. WCIS is a census of all reported workers' compensation claims in California. All insurers, self-insured employers, and state agencies are required to report all claims to the WCIS. Among extensive data reported to WCIS are age, gender, and class code. This allowed us to create the distribution by age and gender of all reported claims in California for the 7 year period.

We then compare the data on expected injury distribution to data on actual injury distribution by calculating a ratio of actual injuries to expected injuries within each cell. A ratio of 1.0 indicates that workers in a particular gender-age range have an actual injury rate equal to the expected injury rate. That is, they are no more or less safe relative to the average of all workers in the same jobs than we expect based on the nature of their occupations and the number of hours they work. A ratio greater than 1.0 indicates workers in the age-gender cell experience more injuries than expected based on risk and exposure. A ratio lower than 1.0 indicates the worker experiences fewer injuries than expected given their occupation/industry and hours worked.

For all data we use CPS person weights.

Results

Table 1 presents the sample sizes and hours worked for the California CPS sample of wage and salary workers for each age-gender cell. Conditional on working, men work, on average, more hours per week than women in all age ranges. For both genders, hours worked increase to a peak in the 35-54 age range and decline thereafter.

Table 1. CPS Sample Distribution by Gender & Age with Average Hours Worked (2002-2008, California, wage and salary workers)

	Male			Female		
Age	N	Hours	Std Error	N	Hours	Std Error
14-17	6,653	19.20	0.155	6,226	16.29	0.128
18-24	10,652	34.53	0.121	10,227	31.13	0.118
25-34	15,600	41.14	0.085	14,834	37.30	0.092
35-44	16,010	42.58	0.089	16,143	36.89	0.097
45-54	14,535	42.72	0.098	14,848	37.82	0.101
55-64	9,604	41.48	0.129	10,241	36.58	0.131
65-74	5,417	35.83	0.215	6,408	30.32	0.199
75-84	3,501	28.80	0.279	5,082	27.04	0.228
85+	663	30.31	0.591	1,366	29.82	0.463

Next we computed the average riskiness of occupations for workers in each gender and age range by calculating an average 2005 manual premium rate for all workers in each cell. This job risk measure is independent of the hours worked, highlighting just how the level of inherent risk in jobs changes over a typical worker's life span. These data are graphed in Figure 1. The figure demonstrates that the men in each age group are, on average, in substantially riskier jobs. The differential ranges from about 25% to 60% higher risk for men over all age ranges. In addition, we see that the job risk declines consistently with age. The migration from higher-risk to lower-risk jobs as workers age may explain much of the early literature's conclusions that older workers have lower injury rates.

Combining these data on risk and exposure, we can answer the question of whether older workers are safer, or just in safer jobs and/or working fewer hours. The first column of Table 2 for each gender gives the expected distribution of injuries by gender and age derived from CPS and the crosswalk of class risk to industry and occupation. The second column under each gender is the actual distribution of injuries as identified for 2002-2008 from reports to the California Division of Workers' Compensation. The third column under each gender is the ratio of actual injuries to expected injuries.

A number of conclusions arise from analysis of Table 2. The injury risk results for men and women differ in surprising ways. For men, after controlling for occupational risk and exposure, injury rates decline for men in all age groups after 18-24, even after controlling for the chaning risk of their occupations and hour worked. This is consistent with earlier studies, but we are able to generalize the result across the full range of age and occupation with controls for the risk of injury in a more precise manner, not available to previous researchers.

On the other hand, the actual rate of injury for women, after controlling for occupational risk and hours of exposure is constant or even increasing for all age groups from 18-24 through 55-64. That is, women between 25 and 64 experience more frequent injuries than predicted by the risk of their occupations and this relative risk may actually be increasing with age. In addition and quite striking, women's injury rates are uniformly and substantially higher than injury rates for men of the same age after controlling for job risk and hours worked.

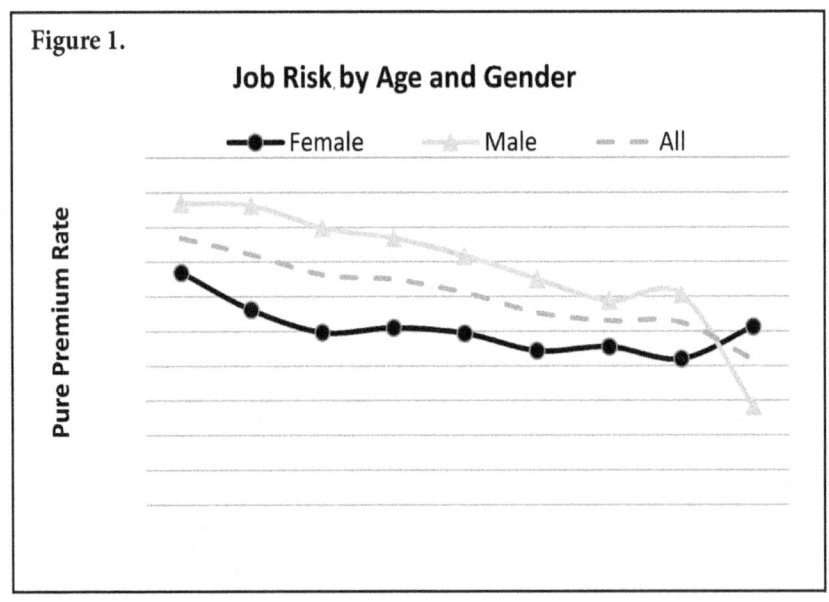

Figure 1.

Discussion

This study highlights a new and important issue. We find that women face a substantially greater risk of occupational injury relative to men when working in similar-risk jobs. Between the ages of 25 and 64, women have injury rates 20% to 50% higher than men working the same number of hours in jobs with similar risk.

This higher risk for women has been missed because women are less likely to be injured overall. Women, despite representing about half the workforce, represent less than 40% of occupational injuries and illnesses. However the overall lower injury rates for women can be attributed to concentration in less risky occupations and fewer hours worked. Once you control for occupational risk and length of exposure, women are much more likely to become injured than men.

Our findings may also explain much of the inconsistency in the conclusions of earlier research on the impact of age on injury incidence. The impact of age will depend heavily on the fraction of the sample that is female.

Over time, women can be expected to continue to increase their participation in higher risk jobs like construction and manufacturing. The higher injury rates for women, when in these jobs, should be a major focus of future research. Experience may be a factor. Male labor force participation is more concentrated at lower ages, meaning, at any age, men are likely to be more experienced. However the constancy of female injury risk over the period 25-64 suggests this is likely to explain only a minority of higher risk for women.

Another explanation could be that higher-risk occupations, traditionally dominated by men, are characterized by workplaces, machinery and safety equipment that is designed for men and poorly adapted for the increasing number of female workers. Future research should focus on the types of injuries and illnesses experienced by women and men in the same occupations. This might highlight the types and consequently the source of the greater risk for women.

Differential reporting, conditional on an injury, between men and women is another explanation that has been put forward. Women may be more likely to report a claim than men, mimicking higher injury rates. The evidence here is inconsistent.

Table 2. Occupational Injury and Illness Distribution by Gender & Age Expected and Actual (California, 2002-2008)

	Male			Female		
Age	Expected	Actual	Ratio (A/E)	Expected	Actual	Ratio (A/E)
14-17	0.84%	0.29%	0.345	0.63%	0.25%	0.397
18-24	9.32%	9.63%	1.033	4.68%	4.92%	1.051
25-34	16.90%	16.13%	0.954	7.29%	8.75%	1.200
35-44	17.15%	15.67%	0.914	8.26%	10.13%	1.226
45-54	13.88%	12.30%	0.886	7.67%	10.18%	1.327
55-64	6.75%	5.48%	0.812	3.90%	4.82%	1.236
65-74	1.44%	0.70%	0.486	0.80%	0.59%	0.738
75-84	0.33%	0.09%	0.273	0.17%	0.08%	0.471

Expected calculated by authors from CPS; Actual tabulated by California Division of Workers' Compensation

Kelsh & Sahl[10] find, after controlling for occupation and hours worked amount electric utility workers, that differential claiming is more pronounced for more severe injuries (inconsistent with reporting bias), but also more pronounced for sprains and strains than traumatic injuries (consistent with reporting bias). However, we cannot discount the possibility that the types of injuries experienced by women, as a result of physiology, are simply different than those experienced by men.

An additional important result that deserves attention in future research is the significant decline in the occupational injury risk for both men and women after the age of 64. For both men and women the decline in the ratio of actual-to-expected injuries after 64 is a substantial break in the previous trend. This is not mirrored by changes in the trend in average risk. An explanation would be a variation on Peek-Asa, et al.[11] healthy worker theory that the least healthy or least able workers are the most likely to retire at or near age 65. However probably the most likely explanation is under-reporting of occupation conditions is considerably more common after workers become near universally eligible for medical insurance under Medicare. This has important implications for the shifting of costs from the private to the public sector.

References

1. Ore, T. Women in the U.S. construction industry: an analysis of fatal occupational injury experience, 1980 to 1992. *American Journal of Industrial Medicine.* 1998;33(3):256-262.

2. Zwerling C, Sprince N, Ryan J, Jones M. Occupational Injuries: comparing the Rates of Male and Female Postal Workers. *American Journal of Epidemiology.* 1993;136(1):46-55.

3. Root N, Daley. Are women safer workers? a new look at the data. *Monthly Labor Review.* 1980; September.

4. Bureau of Labor Statistics (numerous years). http://www.bls.gov/emp/ep_data_labor_force.htm.

5. Toossi, M. Labor force projections to 2016: more workers in their golden years. *Monthly Labor Review.* 2007;130:33-52.

6. Social Security Amendments of 1983 (P.L. 98-21. H.R 1900). http://www.ssa.gov/history/1983amend.html.

7. Society for Occupational and Environmental Health (SOEH) Conference Report. Healthy Aging for a Sustainable Workforce. 2009;November.

8. "California Workers' Compensation Uniform Statistical Reporting Plan" Workers' Compensation Rating Bureau of California. https://wcirbonline.org/wcirb/root/pdf/usrp_ic_regs_only.pdf

9. Neuhauser F, Donovan C. Fraud in workers' compensation payroll reporting: how much employer fraud exists? how are honest employers affected? Report to the Fraud Assessment Commission California Department of Insurance. January 2009.

10. Kelsh M, Sahl J. Sex differences in Work-related Injury Rates among Electric Utility Workers. *American Journal of Epidemiology.* 1998; 143(10).

11. Peek-Asa C, McArthur D, Kraus J. Incidence of acute low-back injury among older workers in a cohort of material handlers. *J Occup Environ Hyg.* 2004;1(8):551-557.

The Mystery of More Monday Soft-Tissue Injury Claims

Richard J. Butler, Ph.D.§, Nathan Kleinman, Ph.D.*, Harold H. Gardner, M.D.*
§Brigham Young University, *HCMS Group

Introduction

All workers' compensation (WC) research examining the temporal distribution of soft-tissue claims—such as low back pain, shoulder, elbow, wrist, hand , knee, ankle, or foot sprains and strains—find more claims filed for injuries nominally occurring on Monday than any other day of the week. Smith (1990) argued that WC creates incentives for workers to report hard-to-diagnose, off-the-job injuries as having occurred on the job. Since there are more off-the-job hours preceding Mondays, and the days after long weekends (referred to collectively as "Mondays") than before regular Tuesdays through Fridays, more off-the-job injuries occur prior to Mondays. Then, hard-to-diagnose injuries will be disproportionately reported on Mondays compared to other regular workdays. Consistent with this hypothesis, Smith showed in WC claims data that a greater proportion of sprains and strains relative to fractures and cuts were reported earlier in the work week and earlier in the work shift than at other work times.

While subsequent research has found, as Smith did, that there are relatively more soft tissue issues on Monday than other days of the week, whether this was due to a moral hazard reporting effect (carrying weekend injuries into work Monday morning because of financial incentives to have the injury treated as a WC claim) has been disputed. Using Minnesota WC claims data, Card and McCall (1996) showed that workers who were less likely to have health insurance coverage were not more likely to report injuries on Monday compared to other days, as would be expected if employees used WC to provide health insurance for non-work related injuries. They also showed that the wage-replacement rate did not exert an independent effect on the probability of Monday injuries. Ruser (1998) found that higher benefits increased the reporting of all injuries on Mondays, but did not raise the probability of a Monday-reported back sprain relative to Monday-reported cut or fracture.

Campolieti and Hyatt (2006) use Canada's universal government-provided medical insurance to identify if the Monday effect was due to health coverage differentials by comparing the Monday effect in Canada with the Monday effect in the United States. Since health insurance coverage for soft tissue injuries is covered under the Canadian health plan, regardless of place of occurrence, it would be expected to find a larger Monday effect in the Unites States than in Canada—if more Monday claims are due to claim migration. Campolieti and Hyatt find quite similar Monday effects in their Ontario sample (Canada) and in their Minnesota sample (United States). They conclude that there results are consistent with an ergonomic explanation of the Monday effect, rather than a moral hazard response.

Methods and Results

Since there is little empirical support for the moral hazard explanation of more soft tissue claims on Monday as an attempt to have weekend injuries compensated through WC insurance as 'mislabeled' Monday work injuries, attention has turned to the ergonomic explanation that after a weekend away from their employment, workers are not sufficiently "warmed up" for workplace tasks and more likely to experience injuries. That is, the workplace is riskier on Monday not because of changes over the course of the workweek in physical risks external to workers, but because workers are simply more likely to experience an injury on Monday because they themselves are different at the beginning of the week because they are "rusty" in the sense that their safety-skills capital depreciates over the weekend.

To address this ergonomic hypothesis, we examine the human resource records of several large firms in a proprietary data base with respect to WC injuries by types (soft tissue vs. other types of injuries), and fatal occupational injuries by day of the week available from the US Vital Statistics. If the ergonomic explanation for more Monday soft-tissue injuries is valid, then we ought to observe more of other types of injuries on Monday, including more occupational fatalities.

While we find about 20 percent more soft-tissue WC claims on Monday than other days of the week, there are actually fewer lacerations/fractures filed on Mondays than other days of the week though the negative Monday response is small and statistically insignificant. Moreover, when we examine the distribution of workplace mortalities nationwide (for 1998, the most recent data available on data ferret), there are fewer fatalities on Monday than other week days:

373 Sunday
719 Monday
848 Tuesday
847 Wednesday
803 Thursday
816 Friday
524 Saturday

If there is something worker's injury risk early in the week, it seems to manifest itself only in complaints about strains and sprains, rather than more widely associated with a fuller range of injuries that would include fractures and lacerations, and more serious workplace risk factors, including workplace mortality. That is, there seems to be no evidence for more general ergonomic risk (external or internal) on Mondays than other days of the week.

Discussion of an Alternative Explanation

Job satisfaction has been found to be a statistically significant explanation of soft tissue injuries. In a prospective study of low back pain, Butler, Johnson, and Cote (2007) find that workers who report at the time of injury to be more satisfied with their jobs subsequently experience less time on a low back claim in the sense of both a lowered likelihood of generating lost time claims and in a lowered likelihood of multiple spells given at least one lost time claim (see also Bigos et al 1991). Gardner and Butler (1996) find in a panel of workers for one company that the likelihood of filing a WC claim subsequently doubles when a worker receives a disciplinary notice (as a measure of dissatisfaction).

If we assume that some subset of workers don't like some aspect of their jobs, and that such workers have a more difficult time returning to work on Mondays (after the weekend away) then they do continuing their work during the week—call this asymmetric work aversion—then such workers would be more dissatisfied with Monday work tend to file more soft tissue claims (this later link is discussed in many of the chapters in Moon and Sauter, 1996). We think this asymmetric work aversion (Monday work aversion) may be reflected in not only reports of soft tissue conditions, but also in absenteeism as well. When we examine the primary ICD9 codes of 840-848 (sprains and strains) in a large proprietary database of employee medical claims (claims made under the employers health insurance policy), we found the following numbers of claims:

353,697 Monday
285,125 Tuesday
338,338 Wednesday
275,903 Thursday
320,730 Friday

Again, more sprains and strains filed by workers generally under their health insurance, though the difference is not nearly as large as the WC differential estimated above.

Using again the proprietary data for several companies that tracked the day of the week the sick leave was taken, the distribution of total hours taken by day of the week are as follows:

93,845	Sunday
2,596,660	Monday
2,195,641	Tuesday
1,998,279	Wednesday
1,961,727	Thursday
2,064,126	Friday
327,246	Saturday

Note how similar are sick hours for Tuesday through Friday, and how remarkably higher the sick hours are for Monday: more than the 20 percent differential on soft tissue filings. Hence, we cannot reject the null hypothesis that the Monday effect is neither ergonomic nor economic, but rather a manifestation of asymmetric work aversion: going back to work is harder than being at work.

References

Butler, Richard J., William G. Johnson, and Pierre Cote (2007) "It Pays To Be Nice: Employer-Worker Relationships and the Management of Back Pain Claims" *Journal of Occupational and Environmental Medicine*, 2007, vol 49, no 2, February; pp 214-225.

Campolieti, M. and Hyatt, D. E (2006). Further evidence on the "Monday effect" in workers' compensation. *Industrial and Labor Relations Review*, **59**(3), 438-450.

Card, D. and McCall, B. P. (1996). Is workers' compensation covering uninsured medical cost? Evidence from the "Monday effect". *Industrial and Labor Relations Review*, **49**(4), 690-706.

Gardner, Harold, and Richard J. Butler (1996) "A Human Capital Perspective For Cumulative Trauma Disorders: Moral Hazard Effects in Disability Compensation Programs," in Sam D, Moon and Steven L. Sauter, editors, (1996) *Psychological Aspects of Musculoskeletal Disorders in Office Work* (Taylor and Francis, London, 1996).

Moon, Sam D., and Steven L. Sauter, editors, (1996) *Psychological Aspects of Musculoskeletal Disorders in Office Work* (Taylor and Francis, London, 1996).

Robertson, L. S. and J. P. Keeve (1983), 'Worker injuries: The effects of workers' compensation and OSHA inspections'. *Journal of Health Politics, Policy and Law* **8**(3), 581–597.

Ruser, J. W. (1998), 'Does WC encourage hard to diagnose injuries?' *The Journal of Risk and Insurance* **65**(1), 101–124.

Ruser, J. W. (1993), 'Workers' compensation and the distribution of occupational injuries'. *The Journal of Human Resources* **28**(3), 593–617.

Smith, R. S. (1990), 'Mostly on Mondays: Is workers' compensation covering off-the-job injuries?'. In: P. S. Borba and D. Appel (eds.): *Benefits, Costs, and Cycles in Workers' Compensation.* (Boston: Kluwer Academic Publishers), 115-127.

Is Occupational Injury Risk Higher at New Firms?

Seth Seabury§, Frank Neuhauser*, John Mendeloff§,#

§RAND Center for Health and Safety in the Workplace, *University of California, Berkeley, #University of Pittsburgh

Abstract

This paper studies whether newly created firms have higher injury rates than established firms. We use data on a large sample of single-establishment firms in Pennsylvania from 2001-2005 to examine the relationship between firm age and the risk of lost workday injuries. Using the full set of firms, there appears to be little overall correlation between firm age and risk. If anything, newer firms appear less likely to have lost workday injuries. When we condition on having at least one injury reported in 2000, however, we find that in later years the injury risk of firms declines with age. This pattern is consistent with systematic underreporting of occupational injuries at newer firms. Surprisingly, we find that firms with a reported workplace injury are less likely to exit within 5 years. More should be done to ensure that occupational health risks are adequately measured in newer firms.

White Papers: Breakout Discussion Group Notes[1,2]

Discussion of: Successes Using Workers' Compensation Data for Health Care Injury Prevention: Surveillance, Design, Costs, and Accuracy.

Michael Hodgson, Lisa Pompeii, Barbara Silverstein, Pat Gucer.
Moderators: Linda Forst and Marie Sweeney

Abstract

Workers' compensation reporting represents an important source of data that complement traditional safety data systems. The healthcare industry has undergone dramatic changes over the last 15 years in terms of data systems development, evaluation of care quality, and public reporting and accountability driven by public awareness and scrutiny. As a result, public report cards, Medicare ranking and comparison reports, The Joint Commission evaluations, and National Provider registries each provide potentially useful information on system performance. Health care, as an example of the service sector, differs from manufacturing and extraction because its "product", people as patients, strongly resembles the work force. This white paper presents implications, strengths, and weaknesses of overlapping existing data systems in three examples, from the State of Washington, Duke University Health System, and the Veterans Health Administration. The structure of the data systems in each system differ dramatically, with linkages variously to human resources (for employment, denominators and rates), to safety data (for job exposure matrices, safety investigation reports, and follow-up), short- and long-term disability costs, patient outcomes, health and wellness, and private health insurance claims data. The resulting structures, constraints, and linkages expand productivity measurement, business case development, safety considerations, and inherent interpretation difficulties. The definitions of events differ substantially between systems; "entry" criteria into systems may differ; the precision and accuracy of operational data often deviate from the expected precision in research settings; and access to data and linkage across systems are governed by competing utilities, laws, and regulations. Workers compensation data can be very useful for identifying and tracking interventions. Nevertheless, users must be aware of the limitations, including under-reporting, selection bias, inaccuracy, and incompleteness. They can be used to evaluate programs, as documented in each of the presentations. Still, business case decisions based solely on workers compensation data cost reductions are likely to miss the added benefits of dramatically improved patient outcomes and other improvements in employee performance. These lessons from health care, where the "production units", i.e., people, resemble the employees, "production" improvement benefits may contribute dramatically to business case justifications, with potential implications for other industry sectors, including education and services. In addition, because of selection bias, strategic decisions based on workers compensation data alone may not reflect the true underlying distribution of risk factors and misdirect goals.

---

[1]Breakout sessions of approximately 45 minutes each were held to discuss the 6 white papers that were drafted for the workshop with two breakout sessions running concurrently at 3 times during the workshop. The discussion notes were collected by session moderators and rapporteurs.

[2]The findings and conclusions of this report are those of the authors and do not necessarily represent the views fo the National Institute for Occupational Safety and Health.

Discussion

Moderators' Question: What are the gaps in the use of WC data in the health care industry? The health care industry is data rich in comparison with many other industries. These resources could be more fully utilized to investigate a range of occupational safety and health issues. A large number of factors and variables in health care industry that could be utilized or investigated further or better delineated in context of or combined with workers' compensation data were discussed. These include:

- Sociodemographic issues: ethnicity, immigration status, literacy level, education
- Length of shift and shift work
- Job change by health care workers after an injury
- What happens to occupations in health care setting such as laundry workers and food service
- Measure safety and reporting cultures
- Assaults and their impacts
- Job descriptions for each work category in health care
- Needlestick injuries
- Loss of earnings and change in occupation due to injuries and illnesses
- Change in hours or change in shift after an injury. Nurses may move to a less strenuous shift to avoid lifting, etc. Other employees change jobs after injury but sometimes there are no jobs for which they are qualified. Some have no computer skills.
- Contracted workers
- Contracted management companies
- Near misses. Not captured in WC or other data. Cannot use the experience of near misses for prevention lessons.
- Home health care
- Exposure hazards not well understood
- Injuries among health care workers but coded in tasks for other occupations by NCCI
- Methodologic approaches for the analysis of data that were recommended by participants:
- Link data WC Claims and First Reports to other data systems
  - Hours of work and shiftwork
  - Incident reports
  - Needlestick injury reports: not comprehensive or consistent across hospitals or systems.
  - Employee health services data
  - Other databases to elucidate relationship between worker safety and patient safety
- Examine incident reports
- Make narratives of injuries more usable and consistent with drop downs/crosswalking schemes
- Make data more real time and accessible to observe effects of changes.
- Utilize electronic filing systems
- Develop a user friendly system to assure accuracy and completeness in coding
- Include data from safety committees
- Track injuries over time
- Create a management protocol for relating disparate systems
- Perform accident investigation and root cause analysis
- Utilization of WC data for prevention and miscellaneous ideas for research and intervention:
- Demonstrate/describe how WC data can be directly applied to initiate interventions
- Describe how WC data can be used to evaluate interventions
- Need better understanding of what WC data captures and what it does not; need better understanding of the contexts in which data are reported
  - Examine "undercount" in WC data utilization
- Explore how to decrease time from reporting to intervention
- Pool data from more than one hospital for comparisons
- Communicate results to "those who need to know"
- Explore how "experience rating" impacts reporting in health care industry
- Study and elucidate the impact of patient safety programs on worker safety

Discussion of: The Total Burden of Work-Related Injuries and Illnesses: A Draft White Paper Developed for the Workshop on the Use of Workers' Compensation Data

Pana-Cryan R, Bushnell TP, Tompa E, Boden LI, Leigh JP, McLeod C.
Moderators: Christine Baker and Rene Pana-Cryan

Abstract

The total burden of work-related injuries and illness is their broad impact on society. This impact extends beyond the number of reported work-related injuries and illnesses and the cost of workers' compensation claims for medical treatment and wage replacement. To operationalize this definition, we would need to assess the total burden accurately but currently there is no one preferred and standardized burden estimation approach.

Estimates of burdens inform decisions that aim to reduce these burdens by comparing them to each other and to strategies to prevent them. It is increasingly important to improve our understanding of the total burden of work-related injuries and illnesses because the pressure continues to build for providing evidence that it pays –at any level, worker's, employer's, or society's– to invest in the safety and health of workers. To articulate this evidence, we need to understand the true magnitude and distribution of the total burden, and as a result, by how much and for whom prevention efforts may reduce it. Currently available burden estimates are being used to make decisions that affect everyone's health-related and economic well-being; improved information is likely to lead to improved decisions.

Despite past efforts to accurately assess the total burden of work-related injuries and illnesses, some of which we mention throughout the paper, gaps remain both at the conceptual and the application levels. Understanding if and how two different burden estimation approaches complement each other is an example of addressing a conceptual gap. Consistently following standardized methods is an example of addressing an application gap.

The primary goal of this paper is to help researchers and consumers of research improve their understanding of the total burden of work-related injuries and illnesses. First, we mention examples of notable studies and present some conceptual relationships among broad estimation approaches and categories of the burden. Then, we elaborate on the difficulties in developing burden estimates that are common in multiple approaches, present criteria for the assessment of the quality of burden estimates derived by different approaches, and briefly describe these approaches, their limitations, and if and how they can utilize workers' compensation data. Finally, we provide recommendations for improving our understanding of the total burden.

Discussion

The total economic and social burden of occupational injuries and illnesses remains uncertain. Many components of the burden are not readily monetized and yet others are spread across society and social support networks. Estimates of total burden would be useful to a number of stakeholders including employers, workers, family members, insurance companies, government organizations, and society as a whole. Various portions of the occupational injuries and illnesses burden are borne by these stakeholders and the costs for each stakeholder are important in the context of the total burden.

Workers and their families often bear many costs that are not recovered from workers' compensation insurance following occupational injuries and illnesses. Governments pay through Social Security and Medicare as well as tax subsidies to workers' compensation insurance programs. Workers' compensation insurance companies may incur costs but these are typically recovered through premiums. Other health insurance coverage may be used for occupational injuries and illnesses and the costs not directly related to the occupational risks. Losses to society occur when, for example, human capital is diminished when workers have to leave their jobs or their occupation following an injury or illness.

Questions and discussions included the following. What kinds of burden estimates are needed for regulatory analysis or for employers? Employers are focused on WC, and are interested in the level of WC costs in comparison with competitors. (However, particularly when all competitors have similar levels of WC costs, these costs can be added by employers to the price of products so that they are borne by consumers). Given the focus on WC costs, it is important to know how WC costs compare to the total burden. Costs for turnover and return-to-work processes may not be directly recognized by employers. Employers may also not recognize the cost savings associated with prevention investments if the impact is on group health insurance and not workers' compensation.

Public health needs to connect with policymakers and employers to show usefulness of WC data and burden estimates, with an emphasis on teaching them how they might use the data. Data systems are generally not integrated. Human Resources and Risk Management don't often share data, so we cannot tell if injury and illness costs are being shifted between WC and group health. Large employers are advocating data integration. Third Party Administrators (TPAs), who need to integrate WC and other data, may be important partners.

There is a need to examine and describe inefficiencies of the WC system. Can we come up with a more complete no-fault system?

Can we avoid costs of fighting (litigation)? More broadly, there are important differences between the Canadian and US systems. One could focus on comparison of Canadian and US burdens to see effects of system on burden.

Estimation
Which portions of the overall burden are small enough to ignore? Where does one draw the line and stop counting indirect costs? For example, if a worker dies on the job, leaving only one parent for their children, the effects on the children could last a lifetime or even be multigenerational. There is a need to define the bounds of an 'episode' which would not likely include impacts on later generations. When there are multiple underlying causes of an illness or injury, assign only part of the cost to the occupational portion.

It is policy at OSHA to use willingness-to-pay measures. (There is current work on a Department of Labor policy paper.) Yet, willingness-to-pay excludes costs to families and employers as well as medical costs since workers do not pay most of them or know what they are. Although popular in the 1980s and 1990s, willingness-to-pay is not used in courts any more. It was hard to understand and apply.

What should we exclude from cost estimates? For example, people may be more at risk from injury or illness at home than at work. Query: Does work injury and illness represent a burden in the relative sense? One answer: Even if the workplace is safer than home, this does not negate the burden of workplace injury and illness, and burden studies inside and outside the occupational health and safety arena do not incorporate these kinds of considerations in their estimates. A fair comparison of burden estimates is therefore facilitated by leaving aside the relative safety of home and workplace.

Data Utilization
Workers' compensation data are useful for identification of cases of occupational injury (although it leaves out claims that are not

filed), some medical costs and legal and administrative costs. Workers' compensation data cannot be used to estimate productivity, human capital, or pain and suffering. Payments to workers are a component of employer costs, but in an economic sense, they are just transfers.

Examination of the linkage between employers' workers' compensation and group health data is important since workers may select to use available group health insurance instead of workers' compensation for compensable occupational injuries and illnesses. Yet, injured worker may not have group health insurance policy through their own employer and may get insurance through a spouse, for example. This leads to a selection bias in the use of group health data.

Medicare and Medicaid have worked aggressively to link their data with WC data in order to avoid covering the costs of cases that should be covered by WC. Some might piggyback on these efforts.

Uncertainties
When using workers' compensation data for burden estimates, one needs to assess the magnitude of the undercounted cases in WC data for claims that are not filed as well as the underestimate of costs per case.

Some health conditions and medical treatments may be partly caused by a workers' compensation injury that occurred many years prior. There is no 'set aside' to cover these costs in the WC system. On the other hand, many age-related decrements are natural or, at least, not occupational. Knees and shoulders, for example, wear out with age, and it is hard to know how much is due to previous occupational injury or work exposure. Longitudinal population studies could be done to see if, generally, WC injuries lead to other conditions down the road, even though it is hard to make the connection between individual WC injuries and later health conditions.

Need to incorporate or build in estimates of indirect costs. Should measures of burden in terms of health-related quality of life be combined with monetary measures?

Summary
Burden estimates need to be designed for a variety of decision makers especially employers and policymakers. Workers' comp data would be most useful for burden estimates when it can be linked to other kinds of data, particularly group health data. An important goal of burden estimates is to identify how and where costs are shifted, as between government and the private sector, and between workers' compensation and group health. Need to focus on how data and burden estimates can be used by employers. Employers need to learn how to integrate WC data with their other data relating to worker health and productivity, and to benchmark their performance against other employers. There are issues related to under and over-estimation of burdens. Some costs may be over-estimated if injuries and illnesses due to both occupational and non-occupational causes are attributed in full to occupational exposures. But we know that most methods of calculating burden, including those based on WC data, represent a systematic underestimate of burden, due to undercounting of cases and due to the omission of some parts of the burden. We need to know the general magnitude of this underestimate in order to make corrections to our burden estimates and make them more useful.

Discussion of: Workers' Compensation Loss Prevention: A White Paper for Discussion.

Joseph Morin, David F. Utterback, Glenn Shor, Len Welsh, Terrance Bogyo, Steven J. Wurzelbacher.
Moderators: Jennifer Wolf-Horejsh and John Mendeloff

Abstract

This paper explores the potential relationship between workers' compensation, loss prevention and public interests by providing a vocabulary and context for discussion. The role, tools and strategies of loss prevention professionals are described, and opportunities for systematic collection and use of available data for analysis, program design and intervention are discussed in detail. Workers' compensation is an important social insurance program with public policy and safety and health objectives that go far beyond the issue of compensation for injury. Loss prevention provisions are explicit in many workers' compensation statutes and implicit in others. Loss prevention services may be applied to assess risk and assist in the insurance underwriting process, or to help employers reduce the human and financial cost of workplace injury. To varying degree and effect, state-sponsored insurers, private insurers and industry safety associations use workers' compensation data for occupational safety and health surveillance and targeting for loss prevention services and initiatives. Several states mandate insurer-based loss prevention services. However, loss prevention data, including employer and insurer-based expenditures, recommendations and interventions, are not systematically collected across industries or jurisdictions. The paper concludes with a discussion of the knowledge gaps and a list of possible collaborations to organize and develop existing loss prevention programs and practices.

Discussion

The insurance industry as a whole makes substantial investments in loss prevention programs. Loss prevention is used to manage and control risks in addition to evaluating risks prior to insuring an employer. The information that is collected for loss prevention programs about hazards, risks, controls, and employer health and safety programs may be useful for identifying intervention needs and evaluating intervention effectiveness. Insurers use loss prevention data for targeted analysis. Good data could be used for predicting injuries.

Issues discussed included the following. Some states require carriers to report loss prevention activities but public reports could not be found. Loss prevention data are usually proprietary to the organization and not standardized. Data owners are unlikely to release the information. State funds might be willing to partner with NIOSH or other groups. AASCIF (American Association of State Compensation Insurance Funds) would be an appropriate industry group to engage.

Loss prevention activities are generally directed at larger employers with higher premiums. What about small employers? Do loss prevention programs evaluate a different segment of work sites than OSHA programs? Work was begun to standardize loss prevention data which would take significant support from the industry. Unclear what may have hindered progress. Standardization could be tried for loss prevention forms and exposure assessment. One could start with occupational health and safety industry groups.

Safety recommendations need to be practically focused so business can implement and see the return on investment. One drawback when talking about loss prevention is that it only applies to employers who use an insurer; it doesn't capture self-insured employers. Is there a way to link loss prevention tools with other inspection information and details?

Discussion of: Contingent Workers: Data Analysis Limitations and Strategies.

Michael Foley, John Ruser, Glenn Shor, Harry Shuford, Eric Sygnatur.
Moderators: Len Welsh and Tim Bushnell

Abstract

The growth of the contingent workforce and workers on alternative work arrangements presents many challenges in the occupational safety and health arena. State and federal laws often entail obligations and rights between employees and employers, but contingent work creates a lack of clarity about who fits into the categories of "employer" and "employee." This results in ambiguities concerning responsibilities to maintain a safe and healthful workplace, difficulties in collecting and reporting data regarding hazards and injuries and illnesses, and raises questions about which, if any, benefit programs are available if health care or disability pay is necessary.

Contingent work may involve uncertainty about the length of employment, control over the labor process, degree of regulatory or statutory protections, and access to benefits. These arrangements introduce specific difficulties in the areas of coverage of occupational injuries and illnesses through workers' compensation, and the particular types and scope of data that the system needs to distinguish between employers' hazard levels and classification.

This paper discusses the various mechanisms under which contingent workers are hired and supervised and explains differences between types of labor contracting, staffing entities, self-employment, and other nonstandard work relationships, including how statistical data sources measure these work categories. It reviews the empirical literature on the relationship between workplace injury and illness risk and different forms of work, including the difficulty of distinguishing the differential impact of temporary work versus employee tenure at work.

The paper highlights differences in availability of regulatory protections and safety net benefits among various types of contingent workers and how these different arrangements incorporate safety and health incentives in the financing of programs. Recognizing the growth in the use of Professional Employee Organizations (PEOs), the paper describes different models for writing workers' compensation insurance policies to cover PEO workers. Finally, it discusses challenges caused by contingent work for accurate data reporting in existing injury and illness surveillance and benefit programs, and opportunities for overcoming obstacles to effectively using workers' compensation data.

Discussion

The phrase "contingent worker" is broad in that it captures groups that may be fundamentally different. It is important to be precise about which type of contingent worker we are considering when discussing or studying this issue. We also need to decide how narrow or broad our target population of concern should be. The term may include:
- Temporary workers
- Professional employment organization (PEO) workers
- Workers employed by a contractor who supplies workers to another organization
- Out-of-country 'labor gangs'

- Misclassified independent contractors without workers comp and other employee protections
- Highly skilled workers such as consultants, free lancers and independent contractors

General characteristics to define contingent work arrangements might include insecurity of employment and expectation that the job will end. Dependence might be considered to be a common characteristic of contingent workers. Perhaps the focus should be 'alternative work arrangements' in general.

There are other categories of workers that are often vulnerable and that may or may not have contingent employment arrangements. However, when they do have contingent work arrangements, they may be especially likely to have higher risk of poor safety and health protections. If contingent work is only one of the sources of vulnerability, or just one of the means by which vulnerability is exploited, then perhaps it is even more important to focus on other sources of vulnerability, or to look at these sources in combination with contingency of work arrangement.

Other categories of vulnerable workers may include:
- Workers who are lower paid
- Non-English speaking
- Less-educated
- Illegal immigrants
- Culturally less oriented toward asserting rights

Some descriptions of contingent workers may not sound respectful to those who work under these conditions.

Understanding the impact of contingent and alternative work arrangements may also require that we understand more fully the various reasons why employers use contract labor providers, PEOs, temporary agencies, etc. Whatever these reasons, insurers are generally supportive of shifts of riskier, more hazardous worker to contractors. While focused on particular alternative employment arrangements, these are part of a more general trend to less secure jobs that is associated with changes in benefits and wages as well as safety.

Perhaps the broader topic of interest is actually 'workplace fissuring' which is part of the more general evolution of the organization of work. This involves fissuring of employer responsibility for workers so that responsibility is divided. Understanding the fissuring of the workplace requires focus on particular industries and how different players participate in the industry's operations rather than on the traditional model of the single, fixed worksite fully representing the industry in which it participates.

One possible goal of gathering data on contingent workers is to determine whether, as a category, they have higher safety and health risks. However, another goal would be enforcement of health and safety and other laws and regulations, which entails identification and tracking of individual employers of concern. Many changes in employment relationships are part of legitimate attempts to re-organize work in more effective ways, but increasing flexibility and complexity in employment relationships also present more opportunities to 'cheat the system,' including attempts to hide or enlarge the use of the underground economy. As employers reorganize, they may take different names and shift workforces out of insured status or transfer them to different policies.

We shouldn't lose sight of the fact that there are good actors representing these new industrial and employment structures (e.g., temporary and PEO agencies) and it's not all about cheating the system. Good citizen businesses, as well as the workers themselves, stand to benefit from strategies that keep contingent workers from slipping through the cracks.

There is a need to keep up with the evolution of organizational changes using new data collection mechanisms and intervention strategies. One basic approach is to attempt to identify each employer in multiple databases, so that they can be better identified and tracked. States may be able to help each other determine how to use multiple databases used for different programs and administrative purposes. However, there are barriers to agencies sharing data – legal as well as operational. While changes in the health care system are increasing attempts to achieve 'interoperability' of data systems, there are also increasing attempts to protect personal identity that can make data linkage and access more difficult. State and federal law changes are complicating the picture.

One tool to use in tracking employers is universal business identification numbers. However, these IDs do not necessarily solve all problems, since employers may change IDs, or have more than one ID. Health care provider reports, such as trauma registries and emergency departments or urgent care clinics can be helpful in identifying workplace injuries when workers' compensation claims are not being filed, especially by workers who are low income, minority, immigrant, etc. A key question is how we can use data to find out why workers are being injured and get interventions implemented to address these causes—and to do this now.

Discussion of: Using Workers' Compensation Administrative Data to Analyze Injury Rates: A Sample Study with the Wisconsin Workers' Compensation Division.

Gregory Krohm, Jennifer Wolf Horejsh, Tracy Aeillo.
Moderators: John Ruser and Terri Schnorr

Abstract

Much of the workers compensation data reported to central state agencies are utilized for administrative and adjudicatory functions. However, codified and narrative information that describe the occupational injury nature, event, source, work process and other factors may be useful for occupational research and surveillance purposes. Each of the 50 states has different rules and regulations that affect the filing of claims and these differences limit comparisons across states. Once a state's system is understood, important relationships between workplace factors and injury patterns can be studied and used for targeting interventions.

Discussion

The breakout group discussed several items that affect the ability to use these data for surveillance and intervention activities.

- Electronic Data Interchange (EDI), a product of IAIABC. Participants in this initiative agree to standardize reporting and coding such as the transmission of Claims, Proof of Coverage, and Medical Bill Payment information through electronic reporting. The most widely used is the First Report of Injury (FROI) submitted to state agencies for all claims.
    - Pros: Every state reports some fields such as nature, body part and cause. By participating, the states are able to reduce paperwork and automate work flow. Multi-state insurers can use the same form and better share knowledge.
    - Cons: Optional fields have variable quality among states, information may be incomplete due to the speed at which they are completed, and the cost of adding variables is substantial.

- Linkages. Linking workers compensation data with other data sets such as wage and hour data may be possible in some states. A common ID would greatly aid in making these linkages. Some states, including Washington, have uniform business ID system but other states may not.

- Potential problems in use of workers compensation data. Some of the many potential limitations of using these data were discussed including how to determine the plant or site of a claim for multi-establishment and multi-industry employers. Some important variables such as occupation, race, and length of service are not included.

- Collaboration of workers compensation agencies with health departments in states. There are many potential opportunities to use these data for prevention. However, public health professionals need to understand the workers compensation system and data and work closely with the industry before considering use of the data for surveillance and intervention purposes.

Discussion of: The Role of Leading Indicators in the Surveillance of Occupational Health and Safety.

Benjamin C. Amick III, Steven Apoodaca, Steven Wurzelbacher.
Moderators: Harry Shuford and Cam Mustard

Abstract

Leading indicators are assessed by collecting information about how the organization is behaving particularly around occupational health and safety. Leading indicators therefore: proceed occupational health and safety outcomes, are a characteristic of the organization, if changed imply occupational health and safety outcomes will change, and implicitly or explicitly incorporate a process of change. The literature in North America has focused on 5 leading indicator concepts: safety culture, safety climate, occupational health and safety management systems, joint health and safety committee and organizational policies and practices. No one metric seems capable of fulfilling all the needs. Thus, there is a need for a broad series of metrics that capture the complexity of organizational and management programs, policies and practices important in predicting injuries, illnesses and work disability outcomes. However key questions remain:

- Many of the questions on these different tools are quite similar, therefore are items in danger of measuring the same construct?
- Which concepts truly predict injuries and illnesses?
- Which questions can be more reliably reported and who is the best in the organization to act as the responder?
- If organizations change do these tools capture the change?
- How do researchers interpret a change in scores using these tools?

Discussion

Consensus was that trailing (lagging) indicators of WC loss data are necessary but not sufficient to rate the effectiveness of OSH programs. For example, focusing on trailing indicators can lead to incentives programs and disciplinary actions that penalize reporting and tend to suppress actual WC rates. Focusing on trailing indicators is limited because it is solely reactive since you are responding to exposures of the past. Also, for small companies, focusing on trailing indicators is not appropriate since most do not have much of a loss experience to focus on. In a similar way, large companies that have been successful in driving down their injury rates need something more to go from "good to great."

Session participants agreed that leading indicators may be useful ("you can't improve a process if you can't measure it"), and focusing on these should be proactive because you can identify system deficiencies that drive future injuries/ illnesses. But the group noted that research in this area has been lacking and that companies do not use leading indicators as much as they should. Most WC research has been done on the claims level – predicting which claims will be the most costly. One participant did question how useful they were, how informative are they? They must present actionable intelligence.

Why are leading indicators not used more by the OSH community? Why has the research not already been done to develop a set of reliable, validated measures developed? The use of leading indicators of one form or another has been increasing. Insurance companies use them every day in assessing exposure/ control for the purpose of risk selection and loss prevention. Many large companies also have developed methods to evaluate the effectiveness of their systems.

But there are reasons why more companies do not use them and research is lacking. First, it is difficult to do and there are many indicators that have been developed but few that have been tested for reliability or validity. Second, the indicators can touch on many levels of the organization and go beyond measuring safety of a task or process. Progress is really a number of small steps forward. For example, how do you measure a concept of "management commitment? Once a measure has been developed, it has to be tested for reliability. Third, there is little evidence for support of leading indicators in terms of their validity and ability to predict injuries/ illnesses. They may well be linked to reduced WC outcomes, but few studies have been designed to show this.

Suggestions for additional research include:
- Define the universe and purpose of indicators
    - Need a more unified field view and a logic model to establish framework for purpose
    - Indicators can range widely across many levels of organization – from Operational- employee, task, job, process, department, site, enterprise to Strategic- organizational policies and procedures, practices, and the value of OSH to the firm; it is difficult to scale data up e.g. translate site level data up to the enterprise level;
    - Exposure vs. control, can measure levels of exposure (e.g. injury potential from cuts, MSD risk factors) or levels of controls in place (safety, ergo, IH)
    - Primary (Safety, ergonomics, industrial hygiene; based on structured safety management systems- OSHA VPP, Z10, AIHA, ISO18001) through secondary/ tertiary prevention (early reporting, disability management)

- Establish reliability of measures- test/retest; internal consistency of scales; among different types of respondents- employees, employers (safety, management, engineers, production), and external consultants, regulators;

- Establish validity of measures to determine what indicators are linked to increased or decreased injuries/ illnesses or WC outcomes
    - Most prior research has been cross sectional
    - More prospective large sample studies needed to determine how large of a difference in outcomes is significant? Which elements of OSH programs are most important? This again may differ based on company size/ industry. Also, the difference should be sustained (over several years) to be noted as significant not just in one year.
    - Outcome is typically WC measures or other trailing indicators- which we know are not perfect so presents problem of measuring validity against questionable standard;
    - Must control for other variables and match performance of companies in like industries

- Establish usability of measures- SMART- specific, measurable, actionable, realistic, and timely and be able to be used across cultures and resonate globally; there may be sets of core indicators that are useful for companies of all sizes and industries; but most indicator may be dependent on the company size/ level of maturity of OSH program, and industry exposures; also upon purpose- (e.g. employer vs. insurance company, regulators). Just because something is predictive does not mean it contains actionable OSH intelligence.

Final Workshop Discussion Group

David F. Utterback, PhD, Teresa M. Schnorr, PhD
National Institute for Occupational Safety and Health

The Use of Workers' Compensation Data for Occupational Safety and Health Workshop concluded with a plenary discussion session. Opportunities and avenues for collaboration and cooperation were discussed among other items. Below are some points that were raised by the workshop participants.

Workers' compensation data can be used to influence leadership within industry and government on the needs for and benefits of occupational safety and health interventions. WC data can also be used for secondary and tertiary prevention. A national database of WC claim information is not available nor is there a representative sample.

Annual summary statistical bulletins and other research reports are published by a number of workers' compensation insurance organizations such as NCCI, IAIABC, NASI and WCRI. No single data source captures all occupational injuries and illnesses so estimates of the injury and illness frequencies and rates may be low even within jurisdictions. Even some major injury types such as concussions, amputations and fractures may be underreported.

Research collaborations are best based on shared interests. Shared usage of data requires long-term commitments and collaboration since one should not anticipate that the meaning and context of data are self-explanatory. Research is frequently completed through collaborative agreements among private and public organizations and institutions. Public agencies have extensive experience with collaborations and public health organizations are familiar with needs and methods for protecting private and confidential information. Prior agreements are informative on ways to overcome concerns about the confidentiality of information. State funds and state workers' compensation agencies may share public interest in injury and disability prevention. Use of WC information for targeting government inspections is not likely to be persuasive to gain collaboration from private organizations.

Some industries such as health care have numerous informative data sets. By comparing occupational injury and illness information from different sources including WC, the limitations of each source can be identified. Such comparison and linking of data can lead to better estimates of the total numbers and rates of occupational injuries and illnesses but this is not a trivial task. For example, the total burden of occupational injuries, illnesses and fatalities cannot be precisely estimated since sources report varying numbers and elements of the burden are not measured in ways that may be readily combined.

Approximately 30 states utilize the IAIABC electronic data interchange (EDI) first reports of injuries. These reports are collected by respective state agencies for portions of their total claims that resulted in loss work time greater than the individual state minimums to qualify for indemnity payments. Some states also aggregate the "medical only" claims. State WC agencies collect and use claims data primarily for performance metrics on the delivery of services. WC data owners may not be authorized to share data with others – even with agencies in their own states in some cases. Except for the standardized data collected by NCCI from

---

Disclaimer: The findings and conclusions in this paper are those of the authors and do not necessarily represent the views of The National Institute for Occupational Safety and Health.

40 jurisdictions, the state level data are probably most complete. NCCI seeks collaboration research with third parties dependent upon the research topics and subject to data limitations and restrictions.

Research by the different interests groups may lead to greater knowledge to reduce injuries and illnesses and show that cost savings are obtainable for interventions. For example, NCCI data might be used in collaborative research to examine patterns of claims for states with innovative loss prevention program requirements. Best practices for the use of WC data to identify intervention needs could be developed. Other projects might compare detailed claims data at NCCI with first report of injury form information from a few states. Loss prevention data that may be useful for research and surveillance could be examined. More standardization of WC data elements and coding systems would increase utility. A primer on WC could be written to address the informational needs of researchers and public health. Workshop participants who are researchers may join the group formed by John Burton. Their annual meetings are very informative and discuss ongoing research projects.

For public employees, WC costs are frequently the only driver to gain the attention of management since OSHA standards do not apply in many jurisdictions. The addition of public sector data to the BLS Survey of Occupational Injuries and Illnesses has been quite informative and reveals greater injuries and illnesses for government employees in many occupations/industry segments than what occurs in the private sector. Public employees also have greater rates of injuries in group health data.

Ideas for future workshop topics were discussed. These included:
- Factors which impede full reporting of injuries by employees and employers
- Linkage of WC with group health data
- Chronicity of disability due to injury or illness
- Leading indicator and metrics to measure effectiveness – best practices should be identified
- Loss prevention methods and metrics from private insurance as well as state funds
- Panel of state legislators on what drives their changes to state requirements
- Practical applications of workers' compensation data for employers and labor

In conclusion, the information shared at this workshop demonstrates the value of utilizing workers' compensation data for occupational injury and illness prevention. Much of the data are collected systematically and longitudinally and maintained in organized databases. Many employers and the insurance industry do use the data to better understand and manage their risks. Public health programs can use WC data to discover intervention needs and evaluate their effectiveness, and improve administration of their occupational health programs. Together, all can help control the substantial occupational injury and illness costs to workers, employers, the economy, and society.

Appendix A
State Health Agencies' Access to State Workers' Compensation Data: Results of an Assessment Conducted by the Council of State and Territorial Epidemiologists, 2012

Letitia Davis, Sc.D.§, Kenneth D. Rosenman, MD#, Glenn Shor, Ph.D.*, Erin Simms, MPH†, Kimberly Miller†

§Massachusetts Department of Public Health, #Michigan State University, *California Department of Industrial Relations, †Council of State and Territorial Epidemiologists

Purpose

The Council of State and Territorial Epidemiologists (CSTE) has a cooperative agreement with the National Institute for Occupational Safety and Health (NIOSH) to help states build capacity to conduct surveillance of work-related illnesses and injuries. One potentially useful source of occupational health surveillance information is Workers' Compensation (WC) claim data[1]. To learn more about whether state public health departments have access to WC data for surveillance purposes and the type of data they have access to, CSTE conducted a brief assessment of health departments in the states. The assessment findings will help CSTE plan future efforts to increase state capacity to use WC claim data for surveillance of work-related injuries and illnesses.

Methods

In March, 2012, a workgroup comprised of representatives from CSTE with input from OSHA developed the CSTE Assessment of State Public Health Department Access to State Workers' Compensation Data. A pilot questionnaire was administered to three states and edited as indicated. The final assessment included 27 individual questions regarding data access, the type of data accessed, the format of the dataset, use of the data, and any barriers or restrictions placed on state WC claims data available to the public health department.

The assessment was made available to all 50 states and the District of Columbia via a Web-based application on April 20, 2012 and remained open for completion for one month. The occupational health contact[2] in each jurisdiction served as the primary informant, although, where applicable, contacts were asked to refer the questionnaire to the person within their department most appropriate for completing the assessment. Thirty-eight jurisdictions completed the assessment. Follow-up was conducted in August 2012 with three jurisdictions to resolve inconsistencies. In this document, 'states' and 'respondent' refers to all 50 states as well as the District of Columbia.

Results

Thirty-eight (75%) state health departments responded to the assessment. Percentages reported below are calculated using the 38 responding states as the denominator.

Access to electronic WC claim data

Eighteen state health departments (47% of respondents) indicated that they had access to an electronic database of WC claim data maintained by the state WC agency[3]. Of the 20 (53%) respondents that did not have access to an electronic database of WC claim data[4]:

- Seven states reported that they had not tried to access the data.
- Three states reported that the WC agency did not maintain electronic records, but they have access to paper records.
- Seven states reported that legal and confidentiality issues were barriers to accessing workers' compensation claim data.
- Two states reported that other agencies maintain data and will provide information upon request;
- One state cited general data access issues as a barrier; and

- One state reported that another agency maintains the data and prefers to work on its own.

Type of WC data accessed
- Five (13%) state health departments reported having access to information about lost wage claims only.
- Twelve (32%) states reported having access to information about both lost wage and medical claim data[5].
- One (3%) state did not know the types of claim data to which they had access.

Types of records accessed
- Thirteen (34%) state health departments reported having access to Employer First Reports of Injury. Among these, four states reported also having access to Physician First Reports.
- Five (13%) states did not know specifically from which records the information received is obtained.

Claims filed vs. claims awarded
- Nine (24%) state health departments reported that they could differentiate between claims filed and claims awarded.
- Six (16%) states reported that they had information on all claims filed but did not have information on whether claims were awarded, pending or denied.
- Three (8%) states reported having data on awarded claims only.

Timeframe
- Two (5%) states reported having daily online access.
- Two (5%) states have weekly access.
- Two (5%) states have monthly access.
- One (3%) state has quarterly access.
- Three (8%) states have annual access.
- Eight (21%) states have access on request only.

One state with weekly and the one with quarterly access have access to claims only for targeted health conditions in these time frames but access to all claims annually. These states were categorized at the most timely point of access.

Several states with routinely scheduled access also reported being able to obtain additional data on request. One state noted that although they receive real time access to self-insured claims, these are not entered into the electronic database.

Data format
- Two (5%) states had online web-based access.
- Eight (21%) states received data through electronic transmission of files to the agency.
- Five (13%) states reported that the data are sent to the public health agency on CD ROM.
- One (3%) state reported access only to paper reports of computer generated files.
- One (3%) state reported actively download data from state mainframe and one (3%) state received the data through a state data clearing house.

Some states reported receiving data in multiple modes. In those instances, these states are categorized by the fastest mode. For example if a state received data both by CD ROM and electronic transmission, this state is included in the electronic transmission category.

- Fourteen (37%) states reported that data can be readily searched and analyzed.
- Four (11%) states reported that while some data can be readily searched and analyzed, other data is in scanned format only.
- No state reported access to data in scanned format only.

Restrictions on data access
- One (3%) state reported having access to claim data only for conditions reportable to public health.
- Seventeen (45%) states reported access to claims for all health conditions. As noted above, in several states, the timeframe for accessing data on targeted conditions was more frequent. One state reported getting personal identifiers for reportable conditions only.

Interagency memoranda of understanding
- Among the 18 state health agencies accessing WC data, 12 have memoranda of understanding with the WC agency, and 6 do not have memoranda of understanding.

Restrictions on data use
- Ten states can use the WC data for aggregate data analysis only. Three of these states reported specifically that they may not use the data for public health follow-up of workers or worksites or for referral to enforcement agencies. One of these reported that they receive neither worker nor employer identifiers.
- Of the remaining eight states, two reported having no restrictions "other than HIPPA requirements" or "as long as appropriate authorities are contacted" and five reported not being able to release the name of the worker either publically or to other agencies; three reported not being able to release the name of the employer publically and two of these could not release the name of the employer to other agencies.

WC agency review of reports
- Four (11%) states reported that they may only publish summary data after review and approval by the WC agency.

Information on dollar amounts awarded to workers
- Six (16%) states have access to information on the dollar amount awarded.
- Ten (26%) states have no information about the dollars awarded.
- Two (11%) states with access to WC data didn't answer the question.

How state public health agencies are using WC data[6]
- Eleven (29%) states use data for Occupational Health Indicators.
- Five (13%) states use data for individual case follow-up.
- Nine (24%) states use data for routine summary data analysis.
- Ten (26%) states use data for periodic special studies of select populations or health conditions.
- Eight (21%) use data in multisource surveillance of targeted conditions
- One state reported using as a source of information for their Bureau of Labor Statistics – Census of Fatal Occupational Injuries program.

Barriers to using WC data for occupational health surveillance
The most highly ranked barrier to using WC data was limited resources, followed closely by "missing key data elements." Data quality issues, and lack of understanding of the data were also relatively highly ranked barriers.

Suggestions for improving use of WC data by state public health agencies
- Seven states (18%) requested assistance in engaging the WC agencies in order to access or maximize use of the WC data. All but one of these seven states were states without access to electronic WC data. Several emphasized the importance of demonstrating a business case and rational for using the data from a public health perspective. "Describing success stories in other states that were able to access and use the data would be useful. "We also must explain the benefit the WC agency would get from sharing the data." Two of the states without access suggested actively educating WC agencies about use of the data for prevention purposes, for example, including WC agencies as well as public health contacts in a webinar on use of the data for prevention.
- Four states (11%), all of which had access to the WC data, reported that additional training on methods and examples of how to analyze and use WC data would be beneficial to their health agency. One state suggested that one-on-one technical assistance would be useful. "We need help in looking at the data and figuring out how to use it."

Table 1. Data elements to which state public health agencies have access in WC data sets

Data element	States with access N	Percent of respondents %
Worker identifiers		
Worker name	10	26
Worker address	9	24
Worker phone number	6	16
Date of birth	16	42
Race/ethnicity	4	11
Preferred spoken language	1	3
Occupation text	10	26
Occupation code	7	18
Bureau of Census (BC)	2	5
Standard Occupational Classification (SOC)	5	13
Both BC and SOC	1	3
Other - unique to WC system	1	3
Employer identifiers		
Employer name	13	34
Employer address	11	29
Employer phone	10	26
Incident location if different than employer address	7	18
Industry description	5	13
Industry code:	14	37
North American Industry Classification System (NAICS)	10	26
Standard Industrial Classification (SIC)	11	29
Both SIC and NAICS	7	18
Injury/illness descriptors		
Nature of injury/illness narrative	7	18
Nature of injury/illness code		
Occupational Injury and Illnesses Classification (OIIC)	6	16
International Association of Industrial Accident Boards and Commissions (IAIABC)	5	13
International Classification of Disease (ICD)	5	13
Workers' Compensation Insurance Organizations (WCIO)	3	8
Other	4	11
Body part descriptors		
Body part narrative	6	16
Body part code:		
OIIC	6	16
IAIABC	0	0
ICD	2	5
Other	5	13
Incident descriptors		
Narrative description of incident	7	18
Date of incident	15	39
Location of incident if different than employer address	8	21
Source narrative	5	13
Source code:		
OIIC	5	13
Other	3	8

- Two states suggested help with standardized coding of the variables, one of which asked for NIOSH coding of industry and occupation. One state suggested seed funding would be helpful for converting their existing system into a fully electronic system, and one additional state would like CSTE to assist them by providing legal guidance on how to overcome confidentiality barriers in WC statutes.

Discussion

Close to half of the responding state public health agencies have access to electronic WC data. This is a significant advance in occupational health surveillance in the states as WC data can be a valuable resource providing critical information for targeting state efforts to prevent work-related injuries and illnesses. State health departments can bring to bear their epidemiologic knowledge and skills to maximize use of this administrative data for surveillance and ultimately prevention purposes.

Notably WC claim data can add substantially to state specific information available from the Survey of Occupational Injuries and Illnesses (SOII). SOII, at the state level, is limited by its relatively small sample size and often cannot provide detailed information by injury or industry characteristics. Also, in contrast to WC data, SOII data currently cannot be aggregated over years.

While assessment findings indicate that state public health agencies can use WC data for population-based surveillance and targeting, restrictions on data use appear to limit its utility for case-based surveillance and case follow-up with either workers or specific employers in many states. Ten of the 18 respondents with access to WC data can use the data for summary data analysis only; two additional states have restrictions on releasing names of employers to other agencies. Thus while the public health agencies in these two states may potentially conduct case or worksite follow-up, they cannot refer cases to enforcement agencies. This leaves only 6 public health agencies that can refer cases for enforcement. However, some state labor departments may have access to WC data for enforcement purposes. This remains to be investigated.

Another notable finding is that a number of state public health agencies do not appear to have sufficient information about the nature, quality and characteristics of the data to which they have access. WC systems are complicated systems, and they vary markedly by state. It is important for epidemiologists using the data to work with the WC agency staff to develop a better understanding of the WC systems in their states and of the data to which they have access. CSTE should provide opportunities for states with more experience using WC data to teach other states about data issues and approaches to maximizing usefulness of WC data for prevention. CSTE should also work with states in developing materials or programs to engage their WC agencies by demonstrating the value of using the WC data for prevention.

Finally it should be noted that there is substantial variation in the types of WC data to which the state public health agencies have access, for example, all claims versus lost time only, awarded claims versus claims filed. These differences in what data can be accessed in different states plus known eligibility differences for WC in different states limits comparison of these data across states. No comparison of data between states should be made without first taking into account what effects differences in access and eligibility have on the data being compared.

Limitations

There are several limitations to this assessment which should be taken into account. First, 13 states, representing 11.5%[7] of the working population in the country, did not complete the survey. If none of the 13 non-responding states had access to WC data, then the percentage of health departments with access would be 35%. If all 13 states had access, then the percentage would be 61%. We suspect that the non-responding states are less likely either to have access or, if they have the legal right to the data, to actually access their state's WC data and the true percentage with access to WC data is closer to 35%. Secondly, only 23 states are funded by NIOSH to conduct occupational injury and illness surveillance that includes use of WC data. The assessment was sent to the individual responsible for the surveillance program in each of these states. In the other 28 states, it is possible the assessment did not reach the person in the state health department knowledgeable about WC data, and therefore the responses from these 28 states may not be accurate.

Acknowledgements

This work was funded in part through a CSTE Cooperative Agreement with the National Institute for Occupational Safety and Health #5R01OH010094.

---

[1] Note that state WC systems do not provide information about injuries and illnesses among workers' covered under other WC systems including federal workers covered under the Federal Employees Compensation Act (FECA), workers covered under the Longshore and Harbor Workers Compensation Act (LHWCA), workers covered by the Black Lung Benefit Program, former nuclear weapons workers covered under the Energy Employees Occupational Illness Compensation Program Act (EEOICPA), and interstate railroad workers covered under the Federal Employers Liability Act (FELA).

[2] CSTE, in collaboration with NIOSH, maintains an updated list of occupational health contacts in state public health agencies. This list is available at: www.cste.org. Washington State is an exception in which the lead contact for occupational health surveillance is in the Department of Labor and Industries.

[3] One additional state health department indicated that it had access to WC claims only for workers in the health department.

[4] States could select multiple answers to this question. Among the 20 states without access to electronic WC data, there were 21 responses from 18 states. Two of the 20 states did not respond to this question.

[5] These 12 included one state that reported "medical claims only" and another that reported access to "First Reports of Injury only" i.e. if state reported access to First Reports of Injury or medical claims only and no additional information was provided, it was assumed that they have access to information about all claims not just those resulting in lost time.

[6] States could select multiple answers to this question.

[7] Obtained from: 1) Bureau of Labor Statistics. Geographic Profile of Employment and Unemployment, 2011. Available at: http://www.bls.gov/opub/gp/pdf/gp11full.pdf (Accessed November 16, 2012). 2) Bureau of Labor Statistics. Employment status of the civilian noninstitutional population, 1941 to date. Available at: http://www.bls.gov/cps/cpsaat01.pdf (Accessed November 16, 2012).

Workshop Participants

Robin Ackerman
OSHA
200 Constitution Ave NW
Washington, DC 20210

Ibraheem Al-Tarawneh
Ohio Bureau of Workers' Compensation
13340 Yarmouth Drive
Pickerington, OH 43147

Benjamin Amick
Institute for Work & Health
481 University Avenue, Suite 800
Toronto, ON M5G 2E9

Karla Armenti
NH Division of Public Health Services
Occupational Health Surveillance Program
29 Hazen Drive
Concord, NH 03301

Christine Baker
CA Dept. of Industrial Relations
1515 Clay Street, 17th Floor
Oakland, CA 94612

Ben Bare
OSHA
200 Constitution Ave NW, Room N3648
Washington, DC 20210

Carrol Bascus
Office of the Assistant Secretary for Policy/ DOL
200 Constitution Ave NW
Washington, DC 20210

Keith Bateman
Property Casualty Insurers Assoc. of America
2600 S. River Road
Des Plaines, IL 60018

Don Bentley
Ohio Bureau of Workers' Compensation
13430 Yarmouth Drive
Pickerington, OH 43147

Deborah Berkowitz
DOL/OSHA
200 Constitution Ave NW, Suite 2315
Washington, DC 20210

Stephen Bertke
CDC NIOSH
4676 Columbia Parkway, MS R-14
Cincinnati, OH 45226

Anasua Bhattacharya
CDC NIOSH
4676 Columbia Parkway, MS C-15
Cincinnati, OH 45226

Elyce Biddle
CDC NIOSH
626 Cochrans Mill Road
Pittsburgh, PA 15236

Leslie Boden
Boston Univ. School of Public Health
715 Albany Street, T-444W
Boston, MA 02118

Terry Bogyo
WorkSafeBC
P.O. Box 5350 Station Terminal
Vancouver, BC V6B 5L5

Paul Bolon
OSHA
200 Constitution Ave NW
Washington, DC 20210

David Bonauto
WA SHARP Program
PO Box 44330
Olympia, WA 98501

Jed Bookman
The Ohio State Univ.
262 Agricultural Engineering Bldg.
590 Woody Hayes Drive
Columbus, OH 43210

Diane Matthew Brown
AFSCME
1625 L Street, NW
Washington, DC 20036

Michael Buchet
OSHA/DOC
200 Constitution Ave NW
Washington, DC 20210

Tim Bushnell
CDC NIOSH
4676 Columbia Parkway, MS R-17
Cincinnati, OH 45226

Richard Butler
Brigham Young Univ.
183 FOB, BYU
Provo, UT 84097

Corey Campbell
CDC NIOSH
PO Box 25226, Denver Federal Center
Denver, CO 80225

Juanita Chalmers
Florida Dept. of Health
4042 Bald Cypress Way, FL 2, Bin A08
Tallahassee, FL 32399

James Collins
CDC NIOSH
1095 Willowdale Road
Morgantown, WV 26505

Amy Coombe
CA Dept. of Industrial Relations
1515 Clay Street, 17th Floor
Oakland, CA 94612

Theodore Courtney
Liberty Mutual Research Institute for Safety
Center for Injury Epidemiology
71 Frankland Road
Hopkinton, MA 01748

Letitia Davis
MA Dept. of Public Health
250 Washington Street, 6th Floor
Boston, MA 02108

Allard Dembe
The Ohio State Univ., College of Public Health
1841 Neil Avenue, 202 Cunz Hall
Columbus, OH 43210

David Elenbaas
MT Dept. of Labor & Industry
PO Box 8011
Helena, MT 59604

Michael Foley
WA Dept. of Labor and Industries, SHARP Program
PO Box 44330
Olympia, WA 98504

Linda Forst
Univ. of Illinois – Chicago School of Public Health
2121 W. Taylor, MC 922
Chicago, IL 606012

Corman Franklin
U.S. Department of Labor
200 Constitution Ave NW, Suite S-2312
Washington, DC 20210

Eric Frumin
Change to Win
110 Williams Street, Room 1201
New York, NY 10038

Abay Getahun
CDC NIOSH
395 E Street, SW
Washington, DC 20201

Patricia Gucer
Occup Health Program, Univ. of MD School of Medicine
11 S. Paca Street, 2nd Floor
Baltimore, MD 21201

Michael Hodgson
Veterans Health Administration
Washington, DC

John Howard
CDC NIOSH
395 E Street, SW
Washington, DC 20201

Jennifer Wolf Horejsh
IAIABC
5610 Medical Circle, Suite 24
Madison, WI 53719

Larry Jackson
CDC NIOSH
1095 Willowdale Road, MS 1808
Morgantown, WV 26505

Bill Kojola
AFL-CIO
Safety and Health Department, 815 16th St., NW
Washington, DC 20006

Gregory Krohm
IAIABC
5610 Medical Circle, Suite 24
Madison, WI 53719

Mike Lampl
Ohio Bureau of Workers' Compensation
13430 Yarmouth Drive
Pickerington, OH 43011

Paul Leigh
Univ. of California, Davis, Dept. of Public Health
One Shields Avenue
Davis, CA 95616

Barry Llewellyn
NCCI, Inc.
111 River Street, Suite 1202
Hoboken, NJ 07030

Margaret Lumia
NJ Dept. of Health and Senior Services
PO Box 369
Trenton, NJ 08625

Chuck McCormick
OSHA/DSG
200 Constitution Ave NW
Washington, DC 20210

Christopher McLeod
The Univ. of British Columbia
201 - 2206 East Mall
Vancouver, BC V6T 1Z3

Eileen McNeely
Harvard School of Public Health
96 Common Street
Belmont, MA 02478

John Mendeloff
RAND Corporation and University of Pittsburgh
3619 Posvar Hall
University of Pittsburgh
Pittsburgh, PA 15260

Nora Ama Mersch
DOL/DEP
200 Constitution Ave NW
Washington, DC 20210

Alysha Meyers
CDC NIOSH
4676 Columbia Parkway, MS R-15
Cincinnati, OH 45226

David Michaels
OSHA
200 Constitution Ave NW
Washington, DC 20210

Laurel Harduar Morano
FL Dept. of Health
605 Jones Ferry Road, #FF03
Carrboro, NC 27510

Alison Morantz
Stanford Univ.
Crown Quadrangle, 559 Nathan Abbott Way
Stanford, CA 94305

Joe Morin
SFM Mutual Insurance Co.
3500 American Blvd W, Suite 700
Bloomington, MN 55431

Cameron Mustard
Institute for Work & Health
481 University Avenue, Suite 800
Toronto, ON M5G 2E9

Nicole Nestoriak
BLS
2 Massachusetts Ave NE
Washington, DC 20212

Frank Neuhauser
Univ. California at Berkeley
2420 Bowditch Street
Berkeley, CA 94720

Stephen Newell
Mercer HSE Networks
1255 23rd St, NW, Suite 500
Washington, DC 20037

Thad Nosal
Insurance Services Organization
545 Washington Blvd
Jersey City, NJ 07310

Josh Novack
DOL/OWCP
200 Constitution Ave NW
Washington, DC 20210

Paul O'Leary
Social Security Administration
Office of Retirement & Disability Policy, 500 E Street, 9th Floor
Washington, DC 20024

Arthur Oleinick
University of Michigan Emeritus
Ann Arbor, MI 48104

Regina Pana-Cryan
CDC NIOSH
395 E Street, SW
Washington, DC 20201

Lyn Penniman
OSHA
200 Constitution Ave NW
Washington, DC 20210

Polly Phipps
BLS
2 Massachusetts Ave NE
Washington, DC 20212

Brooks Pierce
BLS
2 Massachusetts Ave NE, Room 4130
Washington, DC 20212

Lisa Pompeii
Univ. of Texas
1200 Herman Pressler, RAS E617
Houston, TX 77006

Virginia Reno
National Academy of Social Insurance
1776 Mass Ave, NW, Suite 400
Washington, DC 20036

Anthony Robbins
Co-Editor, Journal of Public Health Policy
130 Appleton Street, Unit 1i
Boston, MA 02116

David Robins
Ohio Bureau of Workers' Compensation
13430 Yarmouth Drive
Pickerington, OH 43147

Diane Rodriguez
Veterans Health Administration
Washington, DC

Elizabeth Rogers
BLS
2 Massachusetts Ave NE
Washington, DC 20212

Rachel Roisman
CA Dept. of Public Health
850 Marina Bay Parkway, Building P, 3rd Floor
Richmond, CA 94804

Kenneth Rosenman
Michigan State Univ.
900 Fee Road, Room 117 West Fee
East Lansing, MI 48824

John Ruser
BLS
2 Massachusetts Ave NE, Room 2150
Washington, DC 20212

Terri Schnorr
CDC NIOSH
4676 Columbia Parkway, MS R-12
Cincinnati, OH 45226

Seth Seabury
RAND
1776 Main Street
Santa Monica, CA 90407

Jeanne Sears
Dept. of Health Services, Univ. of Washington
Box 354809
Seattle, WA 98195

Adam Seidner
Travelers Insurance
One Tower Square 7MS
Hartford, CT 06183

Ishita Sengupta
National Academy of Social Insurance
1776 Mass Ave, NW, Suite 400
Washington, DC 20036

John Sestito
CDC NIOSH
4676 Columbia Parkway, MS R-21
Cincinnati, OH 45226

Glenn Shor
OSHA
200 Constitution Ave NW, S-2315
Washington, DC 20210

Harry Shuford
NCCI, Inc.
901 Peninsula Corporate Circle
Boca Raton, FL 33487

Erin Simms
Council of State and Territorial Epidemiologists
2872 Woodcock Blvd., Suite 303
Atlanta, GA 30341

Rosemary Sokas
Georgetown Univ., St. Mary's Hall Room 258
3700 Reservoir Road, NW
Washington, DC 20057

Kerry Souza
CDC NIOSH
395 E Street, SW, Suite 9257
Washington, DC 20201

Emily Spieler
Northeastern Univ. School of Law
400 Huntington Avenue
Boston, MA 02115

Thomas St. Louis
CT Department of Public Health
410 Capitol Avenue, MS#11EOH
Hartford, CT 06134

Robert Steggert
Marriott International, Inc.
10400 Fernwood Road, Dept. 924.36
Bethesda, MD 21701

Garry Steinberg
DOL OWCP
200 Constitution Ave NW
Washington, DC 20210

Eileen Storey
CDC NIOSH
1095 Willowdale Road, MS HG900
Morgantown, WV 26505

Marie Sweeney
CDC NIOSH
4676 Columbia Parkway, MS R-17
Cincinnati, OH 45226

Eric Sygnatur
BLS
2 Massachusetts Ave NE, Suite 3180
Washington, DC 20212

Sangwoo Tak
MA Dept. of Public Health
250 Washington Street, 6th Floor
Boston, MA 02108

Michael Toffel
Harvard Business School
Morgan Hall 497
Boston, MA 02163

Meredith Towle
CO Dept. of Public Health and Environment
4300 Cherry Creek Drive, A-3
Denver, CO 80246

David Utterback
CDC NIOSH
4676 Columbia Parkway, MS R-12
Cincinnati, OH 45226

David Valienti
OSHA
200 Constitution Ave NW
Washington, DC 20210

Lindsey VanderBusch
ND Dept. of Health
2635 E. Main Avenue
Bismarck, ND 58506

Gregory Wagner
CDC NIOSH
395 E Street, SW, Suite 9000
Washington, DC 20201

Renee Walk
WI State Laboratory of Hygiene
2810 Walton Commons W, Suite 200
Madison, WI 53704

Tom Wegener
NYS Workers' Compensation Board
100 Broadway - Menands
Albany, NY 12241

Len Welsh
State Compensation Insurance Fund
333 Bush Street, 8th Floor
San Francisco, CA 94104

William Wiatrowski
BLS
2 Massachusetts Ave NE, Room 4130
Washington, DC 20212

Jeff Wilson
DOL/OASAM
200 Constitution Ave NW
Washington, DC 20210

Steve Wurzelbacher
CDC NIOSH
4676 Columbia Parkway, MS R-14
Cincinnati, OH 45226

Xiaoxi Yao
The Ohio State Univ. College of Public Health
1841 Neil Avenue
Columbus, OH 43210

Theodore Yee
OSHA
200 Constitution Ave NW, MS N3653
Washington, DC 20210

Mark Zak
BLS
2 Massachusetts Ave NE, Room 3180
Washington, DC 21202

Workshop Agenda

Tuesday, June 19, 2012

8:00am - 8:30 am	Check-In	Cesar Chavez Auditorium	
8:30 am - 8:40am	Welcome	Introductions, Facility Safety	David Utterback
8:40am - 8:50 am	Comments	BLS Associate Commissioner for Compensation and Working Conditions	Bill Wiatrowski
8:50 am - 9:50 am	Session 1:	Moderator	Tony Robbins
	Summary: White Paper 1	Successes Using Workers' Comp Data for Health Care Injury Prevention	Michael Hodgson
		The Advantages of Combining Workers' Compensation Data with Other Employee Databases for Surveillance of Occupational Injuries and Illnesses in Hospital Workers	Lisa Pompeii
		Safe Lifting Programs Yield Workers Compensation Savings and Improve Indicators of Resident Well Being	Patricia W. Gucer
		Signature Injuries: Counting and Intervention	Michael Hodgson
9:50 am - 10:10 am		BREAK	
10:10 am - 11:10 am	Session 2	Moderator	Les Boden
	Summary: White Paper 2	Total Burden of Occupational Injuries, Illnesses and Fatalities	Rene Pana-Cryan
		Linking Workers' Compensation Data to Earnings to Study the Economic Impact of Disabling Injuries: Lessons, Challenges and Opportunities	Seth Seabury
		Workers Compensation Costs in Wholesale and Retail Trade	Anasua Bhattacharya

		Linking Workers' Compensation and Group Health Insurance Data to Examine the Impact of Occupational Injury on Workers' and their Family Members' Health Care Use and Costs: Two Case Studies	Abay Asfaw
11:10 am – 11:55 am	Breakout Group	Session 1: Room 1A C5515	
	Breakout Group	Session 2: Room 6 C5320	
11:55 am – 1:05 pm	LUNCH (cafeteria in building, 6th floor)		
1:05 pm – 1:20 pm	Reports: Breakout Sessions 1 and 2	Auditorium	Rapporteurs
1:20 pm – 2:20 pm	Session 3	Moderator	Bill Kojola
	Summary – White Paper 3	Contingent Workers Data Analysis Limitations and Strategies	John Ruser
		Occupational Amputations in Illinois: Report of Data Linkage to Target Intervention Efforts	Linda Forst
		The Scope of PEOs in Employment Markets as Reflected in Workers Compensation Data	Harry Shuford
		The Role of Workers Compensation Data in Occupational Health and Safety Surveillance of Contingent Workers	Mike Foley
2:20 pm –2:30 pm	Comments	Assistant Secretary of Labor for Occupational Safety and Health	David Michaels
2:30 pm –2:50 pm	BREAK		
2:50 pm – 4:00 pm	Session 4	Moderator	Ted Courtney
	Summary – White Paper 4	Loss Control Data Analysis Limitations and Strategies	Glenn Shor
		How WorkSafeBC Uses Workers' Compensation Data for Loss Prevention	Terry Bogyo

		Hitting the Mark: Improving Methods to Target High Hazard Employers Using Workers' Compensation Data	Christine Baker, Amy Coombe
		Injury and Claim Trends in the Ohio Workers' Compensation System	Abe Tarawneh
		Randomized Government Safety Inspections Reduce Worker Injuries with No Detectable Job Loss	Michael Toffel
4:00 pm – 4:45 pm	Breakout Group	Session 3: Room 1A C5515	
	Breakout Group	Session 4: Room 6 C5320	
4:45 pm – 5:00 pm	Reports: Breakout Sessions 3 and 4	Auditorium	Rapporteurs
5:00 pm	Adjourn for day		

Wednesday, June 20, 2012

8:00 am – 8:30 am	Check-In	Cesar Chavez Auditorium	
8:30 am – 9:30 am	POSTER SESSION	See attachment for list of Poster Presentations. Attended by lead author	Room 4215
9:30 am – 10:30 am	Session 5	Moderator	Tish Davis
	Summary: White Paper 5	State-Level Analysis of Workers' Compensation Data for Public Health Purposes	Jennifer Wolf-Horejsh
		Comparison of Data Sources for the Surveillance of Work Injury	Cameron Mustard
		OSHA Recordkeeping Practices and Workers Compensation Claims in Washington; Results from a Survey of Washington BLS Respondents	David Bonauto
		Completeness of Workers' Compensation Data in Identifying Work-Related Injuries	Ken Rosenman
10:30 am – 10:50 am	BREAK		

10:50 am – 11:50 am	Session 6	Moderator	Steve Newell
	Summary: White Paper 6	Leading Indicators and State-level Data Analysis Limitations and Strategies	Ben Amick
		Comparing Injury Data from Administrative and Survey Sources: Methodological Issues	Nicole Nestoriak
		Using O*Net for Studies Relating Psychosocial Characteristics of the Job and Workers' Compensation Claims Experience	Allard Dembe
		Impact of Misclassification on the Estimate of Work-Related Amputation Injuries; Undercount Study of Massachusetts Data: 2007-2008	Sangwoo Tak
11:50 am – 1:00 pm		LUNCH (cafeteria in building, 6th floor)	
1:00 pm – 1:45 pm	Breakout group	Session 5: Room 1A C5515	
	Breakout group	Session 6: Room 6 C5320	
1:45 pm – 2:00 pm	Reports: Breakout Sessions 5 and 6	Auditorium	Rapporteurs
2:00 pm – 3:00 pm	Plenary Discussion	Strategies for future collaborative research and surveillance: Auditorium	
3:00 pm	Adjourn workshop		
	Epilogue	Conversations to continue work	

Poster Presentations

Attended 0830 – 0930, June 20
Room 4215

1. Exploring New Hampshire Workers' Compensation Data for its Utility in Enhancing the State's Occupational Health Surveillance System Karla Armenti, New Hampshire Division of Public Health Services

2. Using workers' compensation data for surveillance of occupational injuries and illnesses – Ohio, 2005 – 2009 Alysha Meyers, Steve Wurzelbacher, Steve Bertke, Mike Lampl, Dave Robins, Jennifer Bell, CDC/NIOSH

3. Use of Workers' Compensation data to enumerate work-related amputations and carpal tunnel syndrome and to determine underreporting of these conditions in California Rachel Roisman, Lauren Joe, Matt Frederick, John Beckman, Martha Jones, David Rempel, Robert Harrison, California Department of Public Health, Occupational Health Branch

4. Describing Agricultural Occupational Injury in Ohio Using Bureau of Workers' Compensation Claims Jed Bookman, S. Dee Jepsen, Mamta Mujumdar and David Robins, Ohio State University

5. Use of multiple data sources to enumerate work-related amputations in Massachusetts: The contribution of Workers' Compensation records L Davis, K Grattan, S. Tak, L Bullock, L. Boden, Massachusetts Department of Public Health

6. Occupational Health Indicators Utilizing Workers' Compensation Data Erin Simms, Kenneth Rosenman, Council of State and Territorial Epidemiologists

7. The Effectiveness of the Safety and Health Achievement Recognition Program (SHARP) in Reducing the Frequency and Cost of Workers' Compensation Claims Ibraheem Tarawneh, Donald Bentley, Michael Lampl and David Robins, Ohio Bureau of Workers' Compensation, Division of Safety & Hygiene

8. Comparison of Cost Valuation Methods for Workers Compensation Data Steve Wurzelbacher, Alysha Meyers, Steve Bertke, Mike Lampl, Dave Robins, Abe Tarawneh, CDC/NIOSH

9. Development and evaluation of an auto-coding model for coding unstructured text data among workers' compensation claims Steve J Bertke, Alysha R Meyers, Steve J Wurzelbacher, Jennifer Bell, Robbins ML, CDC/NIOSH

10. Text Mining for Patterns in Employees' Compensation Appeals Board Decisions Robin Ackerman, JD, SM; Janet Ackerman, BA; Luke Miratrix, PhD; Alex Storer, PhD Occupational Safety and Health Administration

11. Identifying Workers' Compensation as the Expected Payer in Emergency Department Medical Records Larry L. Jackson, Susan J. Derk, Suzanne M. Marsh, Audrey A. Reichard, CDC/NIOSH

12. Utilizing Workers' Compensation Data to Evaluate Interventions and Develop Business Cases James W. Collins, Jennifer L. Bell, CDC/NIOSH/DSR

13. Gender, Age, and Risk of Injury in the Workplace Frank Neuhauser, Anita K. Mathur, Joshua Pines, Center for the Study of Social Insurance, UC Berkeley

14. Higher Monday Work Injury Claims Are More Ergonomic Than Economic Richard J. Butler, Brigham Young University

15. Are Injury Risks Higher at New Firms? Seth Seabury, Frank Neuhauser, John Mendeloff, RAND Center for Health and Safety in the Workplace

www.ingramcontent.com/pod-product-compliance
Lightning Source LLC
Chambersburg PA
CBHW080240180526
45167CB00006B/2358